BIRKHÄUSER

Science Networks . Historical Studies
Founded by Erwin Hiebert and Hans Wußing
Volume 37

Edited by Eberhard Knobloch, Helge Kragh and Erhard Scholz

Helge Kragh

The Moon that Wasn't

The Saga of Venus' Spurious Satellite

with the assistance of Kurt Møller Pedersen

Birkhäuser
Basel · Boston · Berlin

Author:

Helge Kragh
Institute for Science Studies
University of Aarhus
Building 1110
8000 Aarhus
Denmark
e-mail: helge.kragh@ivs.au.dk

Library of Congress Control Number: 2008933401

Bibliographic information published by Die Deutsche Bibliothek
Die Deutsche Bibliothek lists this publication in the Deutsche Nationalbibliografie;
detailed bibliographic data is available in the Internet at http://dnb.ddb.de

ISBN 978-3-7643-8908-6 Birkhäuser Verlag AG, Basel - Boston - Berlin

Basel · Boston · Berlin
P.O. Box 133, CH-4010 Basel, Switzerland
Part of Springer Science+Business Media
Printed on acid-free paper produced from chlorine-free pulp. TCF ∞
Cover illustration: see page 15

ISBN 978-3-7643-8908-6 ISBN 978-3-7643-8909-3 (eBook)

9 8 7 6 5 4 3 2 1 www.birkhauser.ch

To Line, daughter of Venus

Contents

Preface

The planet Venus is the closest neighbour to the Earth and in several respects similar to our globe. It revolves around the Sun at an average distance of 0.72 astronomical units, in an elliptical orbit of eccentricity 0.007. The corresponding numbers for the Earth are 1 and 0.017. The mean density of Venus is 5.2 g/cm^3, that of the Earth 5.5 g/cm^3. Venus' acceleration of gravity at its equator is 8.9 m/s^2, compared with 9.8 m/s^2 at the Earth. The escape velocity is 10.4 km/s, while the corresponding figure of the Earth is 11.2 km/s. Although the mass of Venus is somewhat smaller than that of the Earth – the ratio is $M_V/M_E = 0.815$ – the diameters of the two planets are almost the same. In other words, Venus is indeed a sister planet of the Earth.

In earlier times, when almost nothing was known about the physical conditions of Venus, the similarity appeared even stronger than today. Not only was Venus' period of rotation thought to be close to that of the Earth, it was also possible (and indeed common) to imagine intelligent life on Venus. As late as 1870, the British astronomer Richard Proctor spelled out how remarkably similar Venus was to the Earth: "In size, in situation, and in density, in the figure of her orbit and in the amount of light and heat she receives from the sun, Venus bears a more striking resemblance to the earth than any orb within the solar system. In fact, there is no other pair of planets between which so many analogies can be traced as between Venus and the earth." He then noted: "Had Venus but a moon as the earth has, we might doubt whether, in the whole universe, two orbs exist which are so strikingly similar to each other."[1]

We know today that there is no life on Venus, a planet with most inhospitable physical conditions. Far from being much like the Earth, "Venus is very much like hell," as a Russian astronomer has expressed it.[2] Apart from being lifeless, we also know that Venus, contrary to the Earth, has no appreciable magnetic field. What is here of greater importance, Venus is moonless. It is possible that it once had a

[1] Proctor 1896 (first edition 1870), p. 84.
[2] Shklovskii and Sagan 1966, p. 325.

companion, but if so it disappeared more than a billion years before the emergence of complex life forms on Earth. For this reason, the real moon of Venus, should it ever have existed, belongs to the history of the solar system, not to the history of science.

This is a book about the non-existing satellite of Venus, a fictitious celestial body which played a considerable role in astronomical circles in the eighteenth century in particular. Why write a book about something that manifestly does not exist? The brief answer is that for more than a century the enigmatic satellite – or something taken for it – was occasionally seen, or thought to be seen, and that the object thus became part of the history of astronomy. Although few astronomers believed in the existence of the Venus moon after about 1770, it continued to attract attention for at least another century. It was as if it would not die. By following the discussions of this ghost-like satellite, we address the history of planetary astronomy in a novel way. We get a different insight not only into the world of the astronomers but also into the popular literature concerning the planetary system and other aspects of astronomy.

The book endeavours to give an account of Venus' moon which is as complete as possible. There are undoubtedly details that can and will be filled in by others, but we feel confident that this is a fairly complete history of the subject and that nothing important has been left out. Although it is not the first monograph concerned with the Venus moon – there is a predecessor from 1875, although of a very different kind – it is the first time that the story of the satellite of Venus is told in detail and in its proper historical context. It is a rich and fascinating story, deserving more attention than the few lines to which it is commonly reduced , if mentioned at all, in most books on astronomy and the history of astronomy.

The Moon that Wasn't is based on a large and varied collection of printed sources, published over a period of more than 350 years. Many of these sources are somewhat obscure and have not previously attracted the attention of either astronomers or historians of science. If many of the sources are obscure, so are many of the persons related to the story of the Venus moon. Distinguished astronomers and physicists were certainly involved in both observations and discussions, but they were outnumbered by scientists of a lesser stature, many of them amateur astronomers, not to mention individuals with little or no connection to the astronomical community.

The book is organized in seven chapters, followed by a comprehensive bibliography including nearly 400 primary and secondary sources. After a brief introduction there follows a chapter on the early observation claims, made in the period from 1646 (by Francesco Fontana) to 1740 (by James Short). This chapter also recounts the early history of beliefs in moons accompanying Mars, from Kepler to Voltaire. Chapter 3 focuses on the 1760s, a decade in which Venus passed twice between the Earth and the Sun and therefore offered excellent opportunities for spotting the satellite. Although there were a few claims of having seen it, the majority view was that the Venus transit observations had shown that the satellite did not exist. But then, if the moon does not exist, why had several astronomers of reputation seen it? Attempts to provide an answer were made by Maximilian Hell and Johann Heinrich Lambert, among others, and these are analyzed in Chapter 4. By the end of the eighteenth century the satellite of Venus belonged to history, but it was a history that was much alive. Although Alexander Humboldt dismissed the satellite as nothing but a myth – and he spoke for the majority of scientists – it continued to arouse discussion in astronomical as well as cultural circles. Chapter 5 is largely confined to the period 1790–1860, while Chapter 6 focuses on the 1880s, during which decade interest in the nearly forgotten satellite experienced a kind of revival. The interest manifested itself in scientific commentaries as well as in the science fiction literature of Jules Verne and his contemporaries.

With the extensive analysis of Paul Stroobant in 1887, most astronomers considered the case to be closed and no longer of scientific interest. All the same, the story can be followed well into the twentieth century, which we do in a section of the nature of a postscript. In 1928 a Danish astronomer, Carl Luplau Janssen, gave voice to what had become common knowledge among astronomers. "Venus has no moons," he wrote. "At many occasions in the past it was claimed that a satellite had been seen near Venus, but the discovery has always turned out to rest on a mistake. It can now be considered a fact that Venus does not possess any larger companion." No one contradicted him.[3]

Finally, in the concluding Chapter 7 we summarize the case of the Venus moon and reflect on some of the more general lessons that may be associated with it. Among these are observations of non-existing bodies and the problem of disproving an isolated observation claim. In order to understand the historical course better, we find it valuable to make a comparison between the satellite of Venus and some other objects from the astronomical past, in particular the satellites of Uranus. We also relate the story of Venus' moon to the somewhat similar stories of the Earth's second moon and the discovery history of the satellites of Mars. Following Chapter 7 we include a chapter that comprises a small collection of biographical sketches of some of the central figures in the saga of the Venus moon, from Fontana in the mid-seventeenth century to Stroobant in the late nineteenth century. It is followed by an extensive bibliography, including all relevant primary and secondary sources.

Acknowledgments: We would like to thank librarian Susanne Nørskov for her help with providing sources and Henk Bos, emeritus professor of history of mathematics,

[3]Luplau Janssen 1928, p. 116.

for valuable assistance in understanding and translating Latin texts. Also our thanks to Karin Neidhart, the Birkhäuser Verlag, for her most efficient handling of the book.

List of Figures

Chapter 1

Introduction

The planet Venus is named after the Roman goddess of beauty, love and fertility, the equivalent of Aphrodite in Greek mythology. According to legend she emerged from the foam of the sea onto the island of Cythera (Cyprus), for which reason Venus is sometimes referred to as the Cytherean planet. ("Venerean" or "Aphrodisian" have been proposed, but both names have unfortunate connotations.) Venus may have been the first celestial object clearly recognized as a planet. The Sumerians called it Inanna, the Akkadians chose Ishtar, and to the Egyptians it was known as Astarte. The Chinese knew it as Jin Xing. Venus is the only planet mentioned in the Homeric writings, where it appears in the *Iliad*, announcing the close of day and darkening of night. Achilles' deadly assault on Hektor is described as follows:

> As radiant Hesper [Venus] shines with keener light,
> Far-beaming o'er the silver host of night,
> When all the starry train emblaze the sphere:
> So shone the point of great Achilles' spear.[1]

But Homer and his contemporaries seem to have regarded the planet as two distinct objects, without realizing that the morning star (Phosphoros or Lucifer) is the very same as the evening star (Hesperos or Vesper). The misconception was shared by the Mayas. According to tradition, the insight that the two are merely different manifestations of the same celestial body goes back to Pythagoras. Nor can this insight be found in the Bible, where there is a rare reference to Venus in Isaiah 14:2. It speaks of Lucifer, the light-bearer, and reads in the King James

[1] Alexander Pope's translation of 1715–20, book 22. Available online from `http://ebooks.adelaide.edu.au`

version: "How art thou fallen from heaven, O Lucifer, son of the morning! how art thou cut down to the ground, which didst weaken the nations!"

Of course, the praise of Venus has often been sung by the poets. Among the numerous examples one may single out Lord Alfred Tennyson's poem "Maud" from 1855. The relevant verse is this:

For a breeze of morning moves,
And the planet of Love is on high,
Beginning to faint in the light that she loves
On a bed of daffodil sky,
To faint in the light of the sun she loves
To faint in his light, and to die.

Some years earlier, in 1838, William Wordsworth used Venus to express his worries about the progress of science, which had increased our knowledge of the world but had not resulted in a corresponding moral development:

What strong allurement draws, what spirit guides,
Thee, Vesper! brightening still, as if the nearer
Thou com'st to man's abode the spot grew dearer
Night after night? True is it Nature hides
Her treasures less and less. Man now presides
In power, where once he trembled in his weakness;
Science advances with gigantic strides;
But are we aught enriched in love and meekness?
Aught dost thou see, bright star! of pure and wise
More than in humbler times graced human story;
That makes our hearts more apt to sympathise
With heaven, our souls more fit for future glory,
When earth shall vanish from our closing eyes,
Ere we lie down in our last dormitory?

The painters, too, have found Venus a fascinating celestial object. It appears prominently in Vincent van Gogh's "Starry Night" of June 1889 and also in another of his famous paintings, the "White House at Night" from June 1890, showing a white house at twilight with a prominent yellow star in the sky. According to some astronomers, this star can be identified as Venus as it looked in the evening of 16 June 1890 (art historians do not necessarily agree).[2] But enough of Venus! This is not a book about the planet of love, but of its supposed satellite, a body invisible to the naked eye – indeed invisible to any kind of eye. In a sense, the story of

[2]See *New Scientist* of 28 February 2001.

Venus' moon begins with the chance discovery that the familiar Moon is not the only satellite in the planetary system.

Galileo Galilei created a sensation when he announced in the *Sidereus nuncius* of 1610 his telescopic discovery of four satellites encircling the planet Jupiter.[3] To be precise, "satellite" as a name for a moon was not used by Galileo and only came into general use after it was adopted by Newton in his *Principia*. The name, introduced by Kepler in 1611, derives from the Latin "satelles," meaning an attendant. During the controversy in the seventeenth century over the Copernican system, satellites moving around planets other than the Earth were initially seen as support of the new world system. Galileo realized that the discovery answered a criticism against the heliocentric theory, as it showed that the Earth was not privileged by being the only planet with a companion around it. This was, he wrote, "an excellent and splendid argument" against those who concluded from the presence of our Moon that it constituted a proof of the geocentric world system.[4]

Galileo's discovery was only the beginning of the steady extension of the number of bodies in the planetary system that took place during the seventeenth century. By 1700, ten satellites were known to exist, a number which did not change until William Herschel's discoveries of 1787–89. These discoveries increased the number of satellites to fourteen, and at the beginning of the twentieth century, twenty-five satellites were recognized by the astronomers (see Table 1).

The first and brightest of the Saturn moons (now called Titan) was discovered by Christiaan Huygens in March 1655 and announced the following year in a short publication entitled *De Saturni luna observatio nova.* Two more moons, later known as Iapetus and Rhea, were found in 1671–72 by the Italian-French astronomer Jean Dominique (or Gian Domenico) Cassini a couple of years after he had moved from Bologna to Paris. He originally suggested naming the moons of Saturn *sidera Lodoicea*, after the king of France, Louis XIV. It was also Cassini who discovered Thethys and Dione, which he did in an observation of 1684.[5] Thus, by the end of the seventeenth century, ten satellites were known, one around the Earth, four around Jupiter and five around Saturn. In spite of occasional claims that Mars and Venus might be provided with moons, and also that the number of

[3]Galilei 1989, a modern English translation by Albert van Helden, includes a chapter on the reception of *Sidereus nuncius* (pp. 87–116). Simon Mayer (or Mayr; 1570–1624), a German mathematician and astronomer, claimed to have observed Jupiter's moons since the end of 1609 and in 1614 he published a work entitled *Mundus Iovialis* in which he claimed priority of the discovery. For the priority dispute that followed, see Johnson 1930–31.

[4]Galilei 1989, p. 84.

[5]For contemporary English accounts of Cassini's discoveries, see *Philosophical Transcations* **8** (1673), 5178–5185 and **16** (1686–92), 79–85. Following his discovery of 1684, it took more than a century before more Saturn moons were observed: in 1789 William Herschel discovered two new satellites, now called Enceladus and Mimas.

Planet	**Satellite**	**Distance from planet** (1000 kms)	**Discoverer(s)**	**Year of discovery**
Earth	Moon	384.4	–	–
Mars	Phobos	9.4	A. Hall	1877
	Deimos	23.5	A. Hall	1877
Jupiter	Io	422.0	G. Galilei	1610
	Europa	671.0	G. Galilei	1610
	Ganymede	1 070.0	G. Galilei	1610
	Callisto	1 883.0	G. Galilei	1610
	Amalthea	181.0	E. E. Barnard	1892
	Himalia	11 480	C. D. Perrine	1904
	Elara	11 737	C. D. Perrine	1905
	Pasiphae	23 500	P. Mellote	1908
Saturn	Titan	1 200.0	C. Huygens	1655
	Iapetus	3 558.0	J. D. Cassini	1671
	Rhea	526.8	J. D. Cassini	1672
	Thethys	294.6	J. D. Cassini	1684
	Dione	377.4	J. D. Cassini	1684
	Mimas	185.4	W. Herschel	1789
	Enceladus	238.2	W. Herschel	1789
	Hyperion	1 482.0	W. Bond, W. Lassell	1848
	Phoebe	12 960.0	W. Pickering	1898
Uranus	Titania	438.7	W. Herschel	1787
	Oberon	568.6	W. Herschel	1787
	Ariel	191.8	W. Lassell	1851
	Umbriel	267.3	W. Lassell	1851
Neptune	Triton	353.6	W. Lassell	1846

Table 1.1: The twenty-five satellites known by 1910. The names are the ones adopted by the International Astronomical Union, and the distances are those known today.

moons around Jupiter exceeded four, this number kept stable until the end of the next century.

From the perspective of late-seventeenth-century astronomy the idea of one or more moons around Venus, the planet most resembling the Earth, was in many ways attractive.[6] On the other hand, astronomical or physical theory offered no sure guidance to the question of which planets were equipped with satellites, or how many. Although analogy, pluralism – the belief in extraterrestrial beings – and philosophical expectations of uniformity of nature might speak in favour of a Venus moon, it was realized to be a question entirely left to observation. In his *Dissertatio cum nuncio sidereo*, a work hastily published in 1610 in the wake of Galileo's telescopic discoveries, Johannes Kepler enthusiastically deduced from the Jupiter moons that it followed "with the highest degree of probability that Jupiter is inhabited." Moreover, he suggested that the other planets, too, might possess moons and that the number of these followed a mathematical regularity. He expressed his desire of "discovering two satellites of Mars (as the relationship seems to me to require) and six or eight satellites of Saturn, with one each perhaps for Venus and Mercury."[7] Unfortunately, at the time of writing Kepler did not possess a telescope of the kind Galileo had constructed.

It is of little surprise that the first observation claims appeared in the seventeenth century, as a direct consequence of the use of the telescope, but the further story of Venus' moon is complex and more surprising. Today we know that Venus does not have a moon,[8] but during a period of about 120 years several astronomers observed, independently and at irregular intervals of time, the elusive companion of the planet of love. More than thirty observations are recorded, made by at least twelve astronomers, and there may have been more. For a short period of time the possible existence of the satellite was a real puzzle. To paraphrase a well-known verse:[9]

They seek it here,
they seek it there.
Those savants seek it everywhere.
Is it dark? Or is it bright?
That damned elusive satellite!

[6]For the early temptations to consider Venus and Mars as copies of the Earth, and hence presumably endowed with a moon, see Sheehan 1988, chapters 3–5.

[7]Kepler 1965, p. 40 and p. 77. See further his letter to Galileo of 19 April 1610, in Galileo 1929–39, vol. 10, p. 322. For his Pythagorean numerology, see also Koyré 1961, pp. 138–139. However, such reasoning based on mathematical harmony played little role in the subsequent development of observational astronomy.

[8]Although this is true, it is not quite as true as it was just a few years ago. In 2004 a group of astronomers discovered that Venus has a so-called quasi-satellite (as does the Earth and Mars). This object, called 2002 VE68, is however an asteroid orbiting around the Sun, but in such a way that it appears to travel around the Venusian sky about once every Venus year. See Mikkola et al. 2004.

[9]With due consideration to Leslie Howard, acting as Percy Blakeney in the 1934 movie version of baroness Emmuska Orczy's *The Scarlet Pimpernel*.

However, the latest about 1770, the hope for a moon encircling Venus was effectively abandoned, a main reason being the lack of success in reproducing previous observations. The failure to observe the satellite during Venus' passages over the Sun in 1761 and 1769 was generally taken as proof of its lack of existence. Moreover, most previous discovery claims could be explained as optical illusions, an explanation which enjoyed general support.

The case of Venus' supposed satellite is not well known, not even to modern historians of astronomy. Thus, one looks in vain for it in most accounts of the history of planetary astronomy, and it also does not appear in modern works on the history of the Venus transits.[10] This is not to say that the subject is entirely ignored, for it is sometimes referred to in works on the history of astronomy or in handbooks on planetary astronomy. According to a modern and widely read handbook: "The first report of a satellite of Venus was made by G. D. Cassini on 18 August 1686. Other reports followed, the last being that of Montbaron at Auxerre, on 29 March 1764. It is now certain that no satellite exists, and that the observers were deceived by 'telescopic ghosts.' "[11] However, there is much more to the story, which, strangely, has never been subjected to close and contextual analysis. Most accounts are brief, anecdotal and fragmentary, often uncritically based on earlier writings from the late nineteenth century and rarely on readings of the original sources.[12]

The present work not only aims at remedying this state of affairs by offering a much fuller account based on primary sources; it also follows the story throughout the nineteenth century, at a time when practically no one believed in the existence of the Venus moon. How is it that the non-existing moon continued to attract interest more than a century after its existence was acknowledged as disproved?

[10]Leverington 2003, a detailed history of planetary astronomy from the oldest times to the present, includes no mention of Venus' moon. On the transits of Venus and their relevance for the problem of a possible satellite, see section 3.1.

[11]Moore 1983, p. 61. The statements are not entirely correct. Thus, Cassini's observation was from 28 August 1686. Furthermore, as we shall see, Cassini also reported an observation from 1672, and Fontana claimed to have seen a moon in 1645. The last observation claim dates from 1768, not 1764. William Sheehan, a prolific writer on the history of planetary astronomy, deals only cursorily with the satellite of Venus, which he writes off as nothing but a curiosity. He says, incorrectly, that after 1761 there were no more reports on the phenomenon (Sheehan 1988, p. 24).

[12]The case of Venus' moon is briefly described in Ashbrook 1954, reprinted in Ashbrook 1984, pp. 281–283. For a review by a French astronomer, see Lecomte 1990. More details are given in Hunt and Moore 1982, pp. 92–99, a slightly updated and better documented version of the chapter in Moore 1956, pp. 91–96. Ley 1964, pp. 215–220, draws heavily on Moore's book, but is less reliable. Bakich 2000 only refers briefly to the subject (p. 112). See also `http://www.solarviews.com/eng/hypothet.htm`. Unsurprisingly, much of the information about Venus' satellite found on the internet is unreliable.

Chapter 2

A moon or not? A century of confusion

When Huygens made his observation of Titan revolving around Saturn, a satellite of Venus – meaning something that *might* be a satellite – had already been observed by an astronomer and instrument maker from Naples. However, during the following century the alleged satellite was seen only very rarely. Apart from Fontana's original observation of 1645, it was seen in 1672 and 1686 by Cassini and then in 1740 by James Short in England. That was all. Understandably, at the time when preparations were made to observe the Venus transit across the Sun, predicted to occur on 6 June 1761, the existence of a Venus moon was controversial and enjoyed very little support.

2.1 Venus' satellite observed: Fontana

The first to propose from observations that a satellite might be associated with Venus was the Neapolitan astronomer Francesco Fontana, a lawyer by trade and the most renowned of the early Italian telescope makers. In the 1630s he introduced a new kind of telescope with much greater magnification than the ones used previously.[1] Fontana's telescopes, which made use of two convex ocular lenses, went back to about 1620 and he claimed to have produced the first instrument as early as 1608. He also claimed to have made microscopes at an early date, starting in 1618. By means of his telescope he produced in 1636–38 the first known drawings of Mars, although these were based to a large extent on optical illusions and did not reflect the actual markings of the surface of the planet.

[1]See van Helden 1976a.

Figure 2.1: Engraving of Francesco Fontana (ca. 1580–1656). Reproduced from Hockey 2007, p. 376.

Although appreciated as a maker of telescopes, Fontana's astronomical observations were not held in high esteem. Many of his contemporaries found that his work was superficial and lacking in theoretical arguments, and they tended to dismiss his many reports of celestial novelties. They admitted that he was a fine craftsman, but not more than that. For example, Galileo had nothing but contempt for the Neapolitan telescope maker and rival astronomer who had no training in either mathematics or astronomy. Although he had to admit the greater magnifying power of Fontana's telescopes, he denied that they revealed novelties in the heavens that had not been discovered by his own instruments.[2]

Fontana was well known by his contemporaries, and his name appears frequently in the correspondence of the scientific intelligencer Marin Mersenne and other natural philosophers in the mid-seventeenth century. On 15 November 1638,

[2]See letter of Galileo to unknown correspondent, of 15 January 1639, in Galilei 1929–39, vol. 16, p. 18. See also Winkler and van Helden 1992, pp. 215–216.

a somewhat sceptical Descartes wrote to Mersenne about the new telescopes from Naples: "Do not believe everything you are told about these marvelous Neapolitan glasses; for the majority of people, and the charlatans in particular – such as your Maire, no doubt – always make things they tell of greater than they are."[3] Galileo's former assistant, Evangelista Torricelli, was no less sceptical. In a letter of 1647, he wrote about Fontana's observation claims: "I have the book of foolishnesses observed, or rather dreamed, by Fontana in the heavens. If you want to see insane things, that is, absurdities, fictions, effronteries, and a thousand similar outrages, I will send you the book."[4] Many astronomers were unconvinced that Fontana's claimed observations were authoritative and objective.

The book that Torricelli referred to was Fontana's only printed work, the *Novae coelestium terrestriumque rerum observationes* published in Naples in 1646 and dedicated to Cardinal Camillo Pamphili. A second edition of the book was issued in 1667. Although Fontana stated in the preface that "In 1608 I invented a kind of optic tube," he did his best to flatter the recently deceased Galileo. Referring to the pioneers of telescopic astronomy, he wrote: "As far as the heavens are distant from the Earth, so much do you, Galileo, shine more brightly than the rest." Not a man of modesty, he compared himself to the great Kepler: "Indeed we are both gifted with two talents, to wit, theoretical and practical."[5] A large part of the small book was concerned with his observations of the Moon, which he summarized in several plates. Although he agreed with Galileo that the Moon's light was reflected sunlight, he also attributed some of its light to emanate from the body itself.

In observations of 1636 and 1638 Fontana saw in the middle of Mars "a black cone like a very dark little globule," but was uncertain "whether it was separate from the planet itself and a satellite of it, or rather a big hollow on its disk."[6] He also discovered what appeared to be additional satellites of Jupiter, first in 1630 and later in a couple of other observations: "Five stars, which until now have perhaps been unknown to all astronomers, I have discovered around Jupiter. It can be shown that they are not fixed stars for the reason that fixed stars always keep the same positions to each other, as all astronomers agree. ... Thus I am convinced

[3]Mersenne 1945–88, vol. 8, p. 209. The reference is to Jean Maire (1617–1656), a Flemish librarian and printer. On Mersenne's interest in Fontana and his telescopes, see also vol. 15, pp. 15–16.

[4]Torricelli to Vicenzo Renieri, 25 May 1647, in Torricelli 1919, p. 448. Quoted in van Helden 1994, p. 16. On the sceptical attitude to Fontana's observations, see also Winkler and van Helden 1993, p. 99.

[5]Fontana 2001, p. 15 and p. 26. According to Fontana, he had published lunar observations in 1629 and 1630, but no such works are known to be extant.

[6]Ibid., p. 98.

that these newly discovered bodies are not fixed stars."[7] The claim of nine Jupiter moons had been independently forwarded by the astronomer Schyrlaeus de Rheita in a book of 1643 (see section 2.3), but it is unknown if Fontana were aware of Rheita's claim.

Among Fontana's many observations were markings on Venus, but these were generally believed to be illusory. Unacquainted with Kepler's elliptical orbits, he deduced from his observations that "the orb of Venus is not a perfectly rounded sphere, for if the circle were completed of Venus as seen in those observations, it would not be perfectly round, but an oval shape." In his depictions of Venus, Fontana drew fringes of light around the planet, clearly an optical effect although he did not seem to have recognized it as such. On the evening of 11 November 1645 he observed near the centre of the Venus crescent "a certain spot of a subdued reddish colour," noting that this was "a new discovery, hitherto unknown." He did not report the size of the spot, but from his drawing it appears that it had a radius of about one-fifth of that of Venus. The Neapolitan telescope maker further reported:

> Two small dots were seen to accompany Venus, which I would suppose to be her Courtiers and Attendants, as we shall also call those of Jupiter and Saturn. This is a new discovery not yet published in my opinion. But it is true that they do not always appear, but only when Venus is shimmering, as will be revealed in the diagrams, and these little dots were always seen to be of a more reddish colour. These little dots were, however, not always seen in the same situation on Venus, but they moved back and forth like fish in the sea. From this it can be deduced that Venus itself moves in the same way and is not attached to any part of the sky.[8]

Then, on Christmas day 1645, about an hour after sunset: "There were not two, as in the previous observation, but only one little globe or star seen at the top of the convex side of Venus." And on 22 January 1646: "Venus was discovered to be surrounded by a few rays but with sharper cusps resembling the shape of a bow; a little globe or spot was facing the concave edge of the real Venus."[9] Fontana was apparently uncertain of the nature of the "little globe" he had seen. Although

[7]Ibid., pp. 101–102.

[8]Ibid., p. 91. Note that Fontana speaks in 1646 of a companion of Saturn, a decade before Huygens announced his discovery of Saturn's moon. Fontana undoubtedly referred to the enigmatic appearance of what Galileo had originally called "stars" or "handles" (*ansae*) but which Huygens in 1659, in his *Systema Saturnium*, correctly interpreted as a thin ring surrounding the planet. The last sentence refers to the Aristotelian belief in crystalline spheres that were supposed to carry the planets.

[9]Ibid., pp. 95–96.

he did not unambiguously claim to have discovered a moon moving around Venus, he did speak of his observation as a discovery.

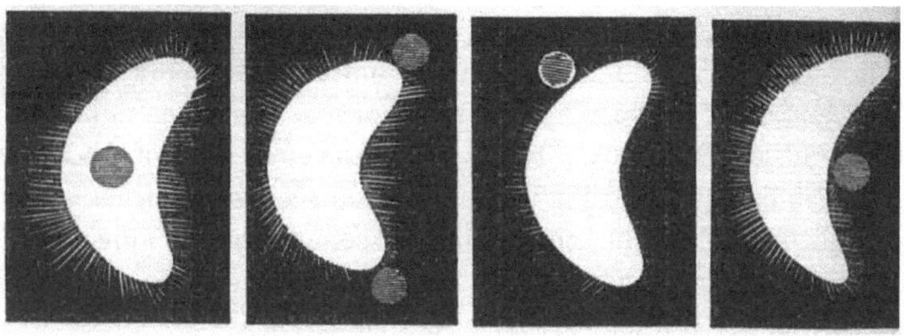

Figure 2.2: Fontana's visual presentations of the supposed satellite(s) of Venus. The figures above are from the woodcuts in Fontana's work from 1646. Below to the left is the redrawing as it appeared in Riccioli's *Almagestum novum* (1651); the figures to the right are from Stroobant's memoir "Études sur le satellite énigmatique de Vénus" (1887).

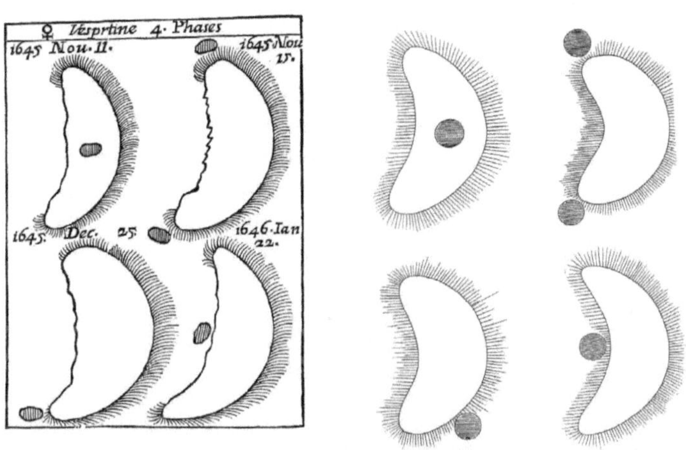

Fontana's observation report quickly caught the attention of the learned world. In a letter of 26 October 1646 the Italian mathematician and natural philosopher Giovanni Battista Baliani reported to Mersenne that he had seen Fontana's book, and he specifically mentioned the claim that Venus appeared to be accompanied by one or two small stars.[10] More importantly, the famous Jesuit astronomer Giambattista Riccioli referred to the claim in his *Almagestum novum* of 1651. Primarily a selenography and an anti-Copernican cosmology, the "new Almagest" also included many of Riccioli's observations made in Bologna. Thus, he described how he had found in 1643 that the unilluminated part of Venus appeared to glow with a strange coppery light, later known as the ashen light.[11] Having described Fontana's observations of the "very ungraceful" sight of a possible satellite of Venus, Riccioli suggested that what had been seen was "either something like a meteor, or a little cloud in the evening, or something like sunspots in front of Venus ..., or the lunar image of a cave and of mountains." He added that "I have never, and neither has [Francesco Maria] Grimaldi nor [Pierre] Gassendi, as seen in Book 3 of his Institutionis astronomicæ, admitted ever to have observed at Venus or close to Venus any globules in any telescope."[12]

A competent astronomer, Gassendi was the first person to report a Mercury transit, such as predicted by Kepler about a year earlier. He witnessed the phenomenon on 7 November 1631 and described it in *Mercurius in Sole visus et Venus invisa*, published in Paris the following year.[13] (Kepler had also predicted a Venus transit to occur on 6 December, but at night-time in Europe; we now know that at Paris it ended half an hour before sunrise.) Gassendi's *Institutionis astronomicæ*, which appeared in 1647, included a reference to the possible moon of Venus, but apparently Gassendi had little faith in Fontana's observations. After having described the appearance of Saturn, he wrote: "Furthermore, it must be added that two [satellites] are carried around Venus, if what is written is true, that two small ones have been observed ... by the Neapolitan Fontana, both of which are reported to have been observed with his own telescope used for [observations of] Mars." Gassendi said that Fontana saw "a little globe at the centre and at the periphery of a circle becoming black," but, he continued, "We have not been able to this day to seize anything about this with our telescope although it was a Galilean one."[14]

[10] Mersenne 1945–88, vol. 14, p. 564. Baliani (1582–1666) corresponded with several of the period's leading natural philosophers, including Galileo.

[11] The elusive ashen light continues to be somewhat of a puzzle as it is observed only occasionally and by some observers. In this respect, it shares some of the features of the Venus moon. On the history of the ashen light, see Hunt and Moore 1982, pp. 87–93, and Baum 2000.

[12] Riccioli 1651, vol. 2, p. 485.

[13] Van Helden 1976b. Although Gassendi (1592–1655) was the only astronomer to publish his observation of the Mercury transit of 1631, it is known that several others saw the phenomenon, including Johannes Baptist Cysat (1588–1657), a student of Christopher Scheiner (1573–1650), and Johannes Remus Quietanus (fl. 1610–1640). The next Mercury transit visible in Europe, on 3 May 1661, was seen by Johannes Hevelius (1611–1687) in Danzig and also by Huygens during a stay in London.

[14] Gassendi 1997, p. 106, a French translation of the Latin edition that appeared in 1658, as

In his important *Selenographia sive Lunae descriptio* of 1647, the Danzig astronomer Johannes Hevelius reported a long series of observations of Venus made in 1644. He accounted carefully for the phases of the planet but saw nothing that could be taken for a satellite. However, he had not seen Fontana's book before he published *Selenographia*, and thus he could not respond to the discovery claim.

According to many later commentators, Fontana's observations were of such poor quality that they could not be taken seriously. They were often dismissed as reflections in his telescope or otherwise due to inexperienced handling of the instrument. To mention but one example, the Austrian nineteenth-century astronomer Karl Ludwig Littrow, director of the Vienna Observatory, thought that "Fontana's telescope was poor and created delusions in the form of illusions." Moreover: "The many rays that he sees around Venus are already most suspicious; he points out explicitly that he sees no satellite when Venus does not 'radiate'."[15]

A prolific author, the famous Jesuit polyhistor Athanasius Kircher published in 1656 a work entitled *Iter extaticum coeleste*, which in its second edition of 1660 became widely known. It included an account of Fontana's discovery claims, as well as a reproduction of Riccioli's drawings of the supposed satellites. Kircher agreed with Riccioli and Gassendi that the evidence in favour of Fontana's moons was less than convincing.[16] Among the early natural philosophers who were aware of Fontana's supposed discovery of a satellite of Venus was also Andreas Tacquet, a Flemish mathematician who taught in the Jesuit colleges of Louvain and Antwerp. In the posthumously published *Astronomia*, included in his *Opera mathematica* from 1669, he mentioned briefly that Fontana had seen two small globes which might possibly be satellites.[17] Referring to the failures of Riccioli, Grimaldi and Gassendi to confirm the observation, Tacquet suggested that it might be due to their telescopes being of inferior quality than the one applied by Fontana, known as an superb artist in telescope making.

One more seventeenth-century natural philosopher who referred to Fontana's satellites was Johann Zahn, an erudite canon of the Premonstrate order in Nuremberg. His *Oculus artificialis teledioptricus sive telescopium* from 1685 was mainly concerned with optical experiments, and the magic lantern in particular, but it also

volume 4 of Gassendi's collected works. The Latin original of 1647 carried the title *Institutio astronomica juxta hypotheseis tam veterum quam Copernici et Tychonis.*

[15]Letter from Littrow (1811–1877) to C. Haase, quoted in Haase 1863–69, p. 254. For similar doubts as to Fontana's observations, see Webb 1868 and Stroobant 1887a, p. 31. Although Bianchini did not believe in the satellite, neither did he believe that Fontana's observations were optical illusions caused by his telescope (see section 2.2).

[16]The edition of 1660 was arranged and edited by Casper Schott (1608–1666), a Jesuit natural philosopher. The discussion of Venus and its possible moons appears in Kircher 1660, pp. 133–136.

[17]Tacquet 1669, part 1, book VII, p. 310. Tacquet (1612–1660) corresponded with Huygens about matters of mathematics. A crater on the Moon is named after him.

included astronomical topics. Among these were a brief discussion of Venus' moon and a redrawing of Fontana's woodcuts of the planet and its strange spots.[18] In his *Specula physico-mathematica-historica notabilium ac mirabilium sciendorum*, a richly illustrated compendium on all contemporary science published in 1696, Zahn adopted a hybrid cosmology between the Ptolemaic and the Tychonian system. His work included a magnificent table showing Venus and its supposed companions (the table was inspired by a similar one in Kircher's *Iter extaticum coeleste*). Apart from referring to Fontana's discovery claim, Zahn also referred to Johannes Wiesel, an optician from Augsburg who may have thought to have observed a moon of Venus. Wiesel was one of Europe's first professional opticians and manufacturers of optical instruments, and his telescopes were in high demand.[19]

The German engineer, natural philosopher and politician Otto von Guericke is best known as an inventor of the air pump and for his spectacular vacuum experiments, but he was also greatly interested in matters of astronomy and cosmology. In his main work, the *Experimenta nova* of 1672, he developed his own system of the world, a Copernican system in which the fixed stars were distributed indefinitely beyond the realm of the planets. Guericke believed that the planets were arranged according to their volume, with the biggest planets farthest away. Contrary to Galileo and most other astronomers, he denied that the sunspots were clouds or spots residing on the Sun, for "we should not postulate the existence of spots or unsightly bodies on the surface of the Sun, the eye of the whole world."[20] Instead he argued that they were small planets moving inside the orbit of Mercury. A similar idea had been proposed sixty years earlier by the Jesuit astronomer Christopher Scheiner, who in a tract entitled *Tres epistolae de maculis solaribus* concluded that the sunspots were actually "stars" or satellites moving very close to the Sun. He even convinced himself that they showed phases like the Moon.[21]

In order to account for the changes and different shapes of the sunspots, Guericke postulated a large number of these intramercurial bodies, all equipped with moons. He seems to have believed that all the planets, intramercurial or not, carried with them one or more satellites. On the other hand, he did not accept the nine moons of Jupiter postulated by Rheita and Fontana.

Acquainted with the works of Rheita, Fontana, Kircher and Riccioli, Guericke supplied Venus with two moons. He had not seen these himself, but relied on the

---

[18]The figure appears on p. 653 of Zahn 1702, a second and enlarged edition published in Nuremberg.

[19]On Wiesel (1583–1662) and his telescopes, see Keil 2000 and Willach 2001. It is unknown if Wiesel, who never published anything, actually made observations of Venus and the other planets.

[20]Guericke 1994, p. 33. *Experimenta nova* also exists in a carefully annotated German edition (Guericke 1968). See also Krafft 2002.

[21]Shea 1970. Scheiner's tract is reprinted in Galilei 1929–39, vol. 5, pp. 39–70.

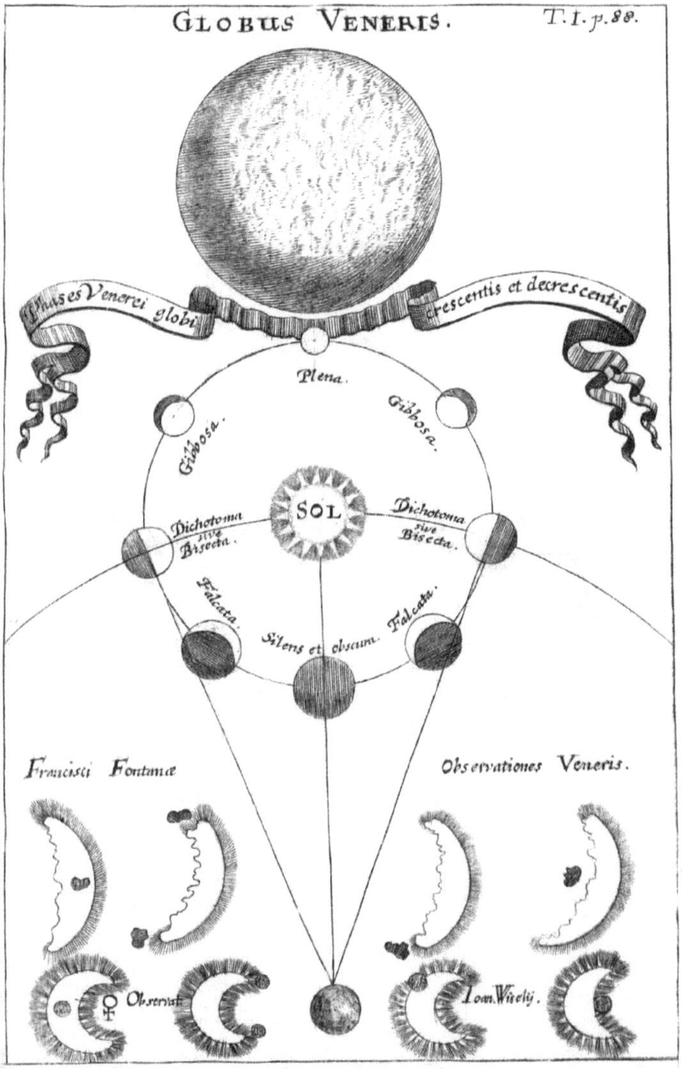

Figure 2.3: Planet Venus in its revolution around the Sun, showing its phases. Also shown, in the lower part of the figure, is Venus' satellite as found by Fontana and Wiesel. Plate from volume I of Johann Zahn's *Specula physico-mathematica-historica* from 1696.

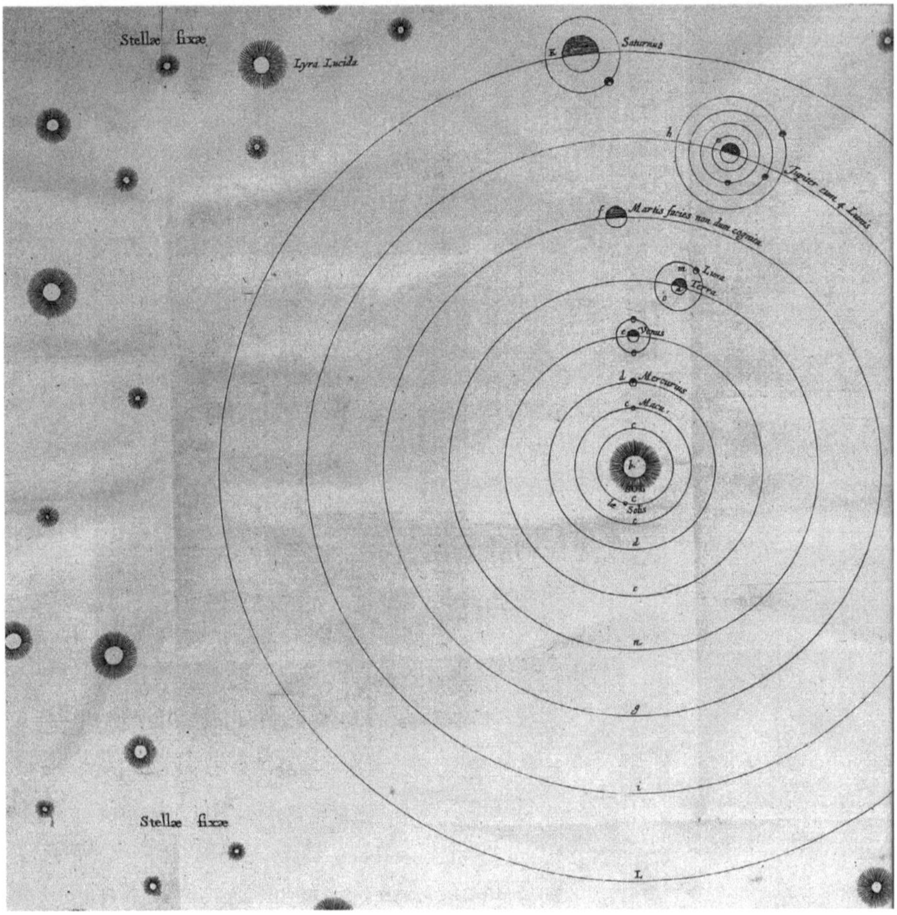

Figure 2.4: Otto von Guericke's system of the world, as included in his *Experimenta nova*, published in Amsterdam in 1672. The system differs from the traditional Copernican system by the absence of a stellar sphere; the fixed stars are spread throughout an indefinitely great space. Also notice the intramercurial planets and the two satellites of Venus. Guericke's system was based on theory, not on observations.

observations of Fontana, who in his second observation "saw a small sphere, dark purple in colour, in the middle of its disk. His third observation found two separate stars of the same dark purple colour adjacent to it, as it were, in the horns of the planet."[22] *Experimenta nova* contained a picture of Venus with two satellites encircling it in the same orbit. Guericke's advocacy of the satellite of Venus was clear but uninfluential. It was ignored in the later literature on the subject.

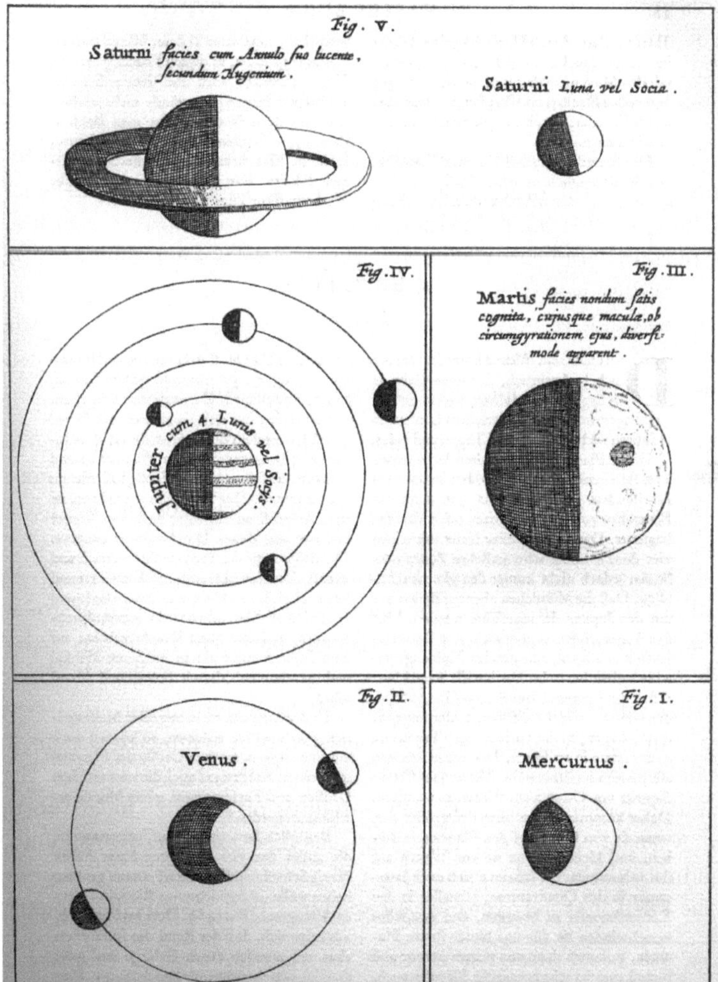

Figure 2.5: The six planets and their satellites, as shown in *Experimenta nova*. The new feature is the two satellites of Venus, moving in the same orbit.

[22] Guericke 1994, p. 36.

Huygens looked with a 12-foot telescope for satellites around the planets Venus, Mars and Saturn. On 23 June 1656, his friend Jean Chapelain told him: "Mr. de Montmor felt obliged by the manner you have written to me that you let him see with your long and excellent telescope if Venus should not have a satellite which turns around her; if Mars has on its disk a kind of pyramid-shaped mountain at the foot of which there is a deep abyss; and also if there should be a planet around him [Mars] analogous to our Moon."[23] Having established the existence of a moon revolving around Saturn, he wrote to Chapelain:

> It therefore seems reasonable that Venus and Mars should be accompanied as the other planets ... and yet I have not seen them [the satellites] to this hour. Furthermore, I have even observed with a 24-foot telescope, but not Venus, so perhaps there is something left around her to be discovered. I have read by Mr. Fontana that he had noticed the three other companions of Venus, but I find it difficult to believe, because he ought then also to have seen Saturn's moons, and because I know that the telescopes he used were only 6 feet long.[24]

At this point Huygens was optimistic. He thought he could discover a satellite around Venus and continued his observations, but after three more years of observation he was forced to conclude that Venus was not endowed with a companion.

In his book on the *Systema Saturnium* from 1659, he came up with some theoretical arguments in support of the conclusion.[25] Although astronomical numerology was not highly regarded in the seventeenth century, it was such reasoning that made Huygens conclude that there were no more satellites to be found. He reasoned that, after his discovery in 1655 of the Saturn moon, the system of the world was numerically complete as it consisted of a symmetrical pair of the first perfect number, six, namely, six planets and six satellites. Therefore, there would be no point in looking for either new satellites or planets. In his posthumously published *Kosmotheoros* from 1698, Huygens discussed the five satellites of Saturn and admitted that "there may be a Sixth; or perhaps there may be another without the Fifth that may yet have escaped us." As to Venus, he thought it "must

[23] Huygens 1888, p. 437. Habert de Montmort (ca. 1600–1679) was a French patron of science after whom the "Montmort Academy" was named. This academy consisted of a circle of natural philosophers who met regularly in Paris between 1653 and 1664.

[24] Huygens 1888, p. 472. Letter of July 1656. Huygens' reference to "the three other companions of Venus" is puzzling, as Fontana made no such observation claim. As to Saturn's moons, Fontana detected "sometimes two satellites, and at most three." He did not identify the ring system, but like earlier observers he described it as two "crescent-shaped objects" or "kinds of handles." Fontana 2001, p. 111.

[25] Huygens 1925, pp. 212–214.

be acknowledged inferiour to the Earth" because it is smaller and lacks a moon.[26]

Of course, the original argument collapsed when Cassini in October 1671 found another satellite around Saturn. Huygens was present, together with Jean Picard, Ole Rømer and other Paris astronomers, and, to quote I. Bernard Cohen, "One can easily imagine Huygens' chagrin on learning that his own numerological inhibitions may have prevented him from discovering more than one satellite of Saturn."[27] To our knowledge, Huygens thereafter never returned to numerology as a scientific argument. Cohen takes a bolder and not entirely justified step: When in 1673 Cassini published his findings of the two new Saturn satellites, that "event is also of interest as apparently the final stage in the influence of numerology on astronomical thought and discovery."[28]

2.2 Jean Dominique Cassini

In spite of the early works of Fontana, Huygens, Guericke and others, it was only with the observations of Cassini that the Venus moon truly made its entry into the astronomical discourse. Born in Perinaldo in Liguria, as a young man Cassini was invited to Bologna, where he studied under Riccioli and Grimaldi. One may assume that he was acquainted with Fontana's book of 1646, including its claim of one or two companions revolving around Venus. At any rate, as a professor of astronomy at the University of Bologna, he was much occupied with observations of Venus. Thus, in 1667 he announced that he had found spots on the planet from which followed a period of rotation somewhat less than 24 hours, a value his son Jacques later revised to 23 hours 20 minutes.[29]

Two years later Cassini was called to Paris to take up a new and highly prestigious position at the recently founded Académie Royale des Sciences and to serve as the first director of the new Observatoire de Paris.[30] During his long and fruitful career in Paris he developed into what a modern historian has called Europe's

[26]Huygens 1722, p. 114 and p. 112, and Huygens 1944, p. 774.

[27]Cohen 1990, p. 204. On Huygens' argument, see also Cohen 1978.

[28]Cohen 1990, p. 199. The suggestion is hardly justified as there are several cases of numerology in the later history of astronomy. To mention but one example, the Titius-Bode law of the distances of the planets, proposed in the second half of the eighteenth century, was certainly numerological in nature. As we shall see in Chapter 6, numerological reasoning was alive and well in the late nineteenth century. Nor, for that matter, is it absent from modern astronomy and cosmology.

[29]Cassini 1735; Cassini 1740, p. 525. In 1877 Giovanni Schiaparelli (1835–1910) suggested a period of rotation of about 225 days. Modern observations tell us that Venus' time of rotation at its equator is 243.019 days retrograde, and its sidereal revolution around the Sun is 224.70 days. For a historical account of the vexed question, see Hunt and Moore 1982, pp. 36–51.

[30]On Cassini and the new observatory in Paris, see Wolf 1902.

finest telescopic astronomer.[31] He made numerous observations of Jupiter's satellites, in 1675 he discovered the gap in Saturn's ring system known later as Cassini's division, and in 1679 he published a new map of the Moon. There is no doubt that he held a very high authority and that his observations were taken very seriously. When it came to theory, matters were different, for Cassini was a conservative who not only opposed Newtonian gravitation but also the finite velocity of light discovered by Ole Rømer in 1676. But this is of no great relevance in the present context.

Figure 2.6: The Paris Observatory in the beginning of the eighteenth century, at a time when J. D. Cassini was still active. Reproduced from Wolf 1902, plate XI.

Cassini never recovered the spots that he had seen before he came to Paris. On the other hand, he observed Venus' moon twice, in 1672 and 1686, in both cases as a faint object showing phases similar to those of Venus. In between, in 1683, Cassini had cooperated with Niccolo Fatio to make pioneering observations of the zodiacal light, a faint glow in the night sky which at the time was thought to be an extension of the solar atmosphere (but is now known to be due to the reflection of sunlight on cosmic dust particles). Details of the events of 1672 and 1686 only appeared in 1730, when his paper on the zodiacal light was reprinted, now with the following inserted:

[31] This is the expression of van Helden 1994, p. 28. According to van Helden, "Cassini's personal prestige, his position at the French Observatoire Royale and Académie Royale, and his access to the best telescopes in Europe made his discoveries accepted by everyone" (ibid.).

> 1686, August 28th, at 4.15 in the morning. Looking at Venus with a telescope of 34 feet focal length, I saw at a distance of 3/5 of her diameter, eastward, a luminous appearance, of a shape not well defined, that seemed to have the same phase as Venus, which was then gibbous on the western side. The diameter of this object was nearly one quarter that of Venus. I observed it attentively for 15 minutes, and having left off looking at it for four or five minutes, I saw it no more; but daylight was by then well advanced. I had seen a like phenomenon, which resembled the phase of Venus, on 1672, January 25, from 6.52 in the morning to 7.02, when the brightness of the twilight caused it to disappear. Venus was then horned, and this object, which was of diameter almost one quarter that of Venus, was of the same shape. It was distant from the southern horn of Venus a diameter of Venus on the western side. In these two observations, I was in doubt whether it was or was not a satellite of Venus, of such a consistence as not to be very well fitted to reflect the light of the Sun, and which in magnitude bore nearly the same proportion to Venus as the Moon does to the Earth, being at the same distance from the Sun and Earth as was Venus, the phases of which it resembled. ... But in spite of some research I have done from time to time after these two observations, in order to complete a discovery of such great importance, I have never succeeded to see it except these two times; and this is why I suspend my judgment.[32]

Like Fontana (to whom he did not refer), Cassini seems to have been uncertain about the nature of the observed object. Although he may have thought that he had discovered a Venus moon, he did not claim so explicitly. Nonetheless, during the eighteenth century it became common to refer to Cassini's observations as constituting either a discovery or a discovery claim.

For more than half a century after Cassini had first sighted the supposed satellite of Venus, no further observations of the moon were made. David Gregory, the Savilian Professor of Astronomy at Oxford, wrote in 1702 an introduction to Newtonian astronomy, entitled *Astronomiae physicae et geometricae elementa*, in which he approvingly referred to Cassini's observations of 1672 and 1686. The

[32] Cassini 1730, p. 245, also quoted in Mairan 1764, p. 162. See also the translation in Hunt and Moore 1982, pp. 94–95, where the date is incorrectly given as August 18, a date that also appears in Cassini 1730, p. 246. The original paper on the zodiacal light, entitled "Découverte de la lumière céleste qui paroist dans le zodiaque," was published in Paris in 1685, i.e., a year before Cassini believed for the second time to have observed the satellite of Venus. The treatise including the section on the observations of the Venus moon appeared twice in 1730. One of the versions was the *Oeuvres diverses de M. I. D. Cassini*, published as a special volume of the *Mémoires* which also included other of Cassini's works. The only difference between the two versions is the pagination (which in the latter case is pp. 121–213).

later English edition included a more elaborate discussion of Cassini's results. Gregory believed that they gave "more than a bare Suspicion to incline us to believe that Venus has a Satellite," and added: "If Venus has its Moon, it is almost in all Respects like our Earth, from which it differs less in the Quantity of its Day and Night, and Degree of Heat, than any other Planet."[33] As to why Cassini had been unable to see the satellite again ("tho' he much endeavour'd it"), Gregory suggested that it reflected the Sun's light much more poorly than the planet. Perhaps the satellite was as poor a reflector as the darkest spots of the Moon, which might be explained by supposing that Venus, but not its satellite, was covered by a nebulous and highly reflective atmosphere.

Gregory did not refer to the earlier observations of Fontana, an oversight which was not repeated by the erudite Roman historian and polymath Francesco Bianchini, who worked for Pope Benedict XII and was a specialist in calendar studies. In 1728 he published *Hesperi et phosphori nova phaenomena*, the first book ever to be written about the planet Venus. Bianchini described in detail his discovery – or discovery claim – of patches and other markings observed on Venus and also his determination of the planet's period of rotation, for which he got a value entirely different from the one obtained by Cassini: Whereas the latter had found 23–24 hours, Bianchini concluded that the planet turned around itself in 24 days 8 hours.[34] Equipped with one of the excellent telescopes of the Roman telescope maker Guiseppe Campani[35] – whose telescopes were also used by Cassini – he thought to have identified several Venus "continents" and "oceans" which he proposed to name after Portuguese and Italian celebrities (his oceans included a *mare Columbi*, a *mare Vespucci* and a *mare Galilæi*).

With regard to Fontana's observations of what was possibly a satellite, Bianchini admitted that there was doubt as to "whether it might be a satellite of Venus, or a meteor in its atmosphere, or another opaque body between Venus and the observer's eye." But he denied that the observations were just caused by errors of the lenses:

> There are some who think that there were some spots in the glass of Fontana's lenses; but one cannot easily suspect this in the case of the learned gentleman, neither should we presume an astronomer to be so ignorant of his specialty that he does not know how to detect this error by rotating the tubes about their own axis, since if it were in the glass

33 Gregory 1702, p. 472; Gregory 1736, vol. 2, pp. 834–835.

34 Bianchini's book exists in an English translation as Bianchini 1996. For details and context, see Dal Prete 2005. On Bianchini and his role in astronomy in the early eighteenth century, see Heilbron 2005 and also Heilbron 1999, especially pp. 148–155, 253–255. The attempts to map Venus, from Fontana over Bianchini to the space age, are surveyed in Moore 1985.

35 On Campani (1635–1715) and his development of optical instruments, see Bedini 1961.

it would at once change its position and it would likewise be projected onto other planets viewed with the same tube [or eyepiece].[36]

Figure 2.7: Bianchini mounting a telescope. Drawing made on 28 September 1720, reproduced from V. Kockel and B. Sölch, *Francesco Bianchini (1662–1729) und die europäische gelehrte Welt um 1700* (Berlin: Akademie Verlag, 2005), p. 21.

Probably relying on Gregory's mention of Cassini's observations, Bianchini reflected on the possible existence of a satellite of Venus and whether or not Cassini had seen the same object as Fontana. He found it most unlikely that the hypothetical satellite had only made itself clearly visible at such rare occasions and therefore concluded that "at this time it seems rather indecorous for an astronomer to want to crowd the side of Venus with a satellite." Rather than ascribing the observations of "Fontana & Cassini" to the atmosphere of Venus, he preferred to explain them in terms of a thickening in the past of "the heavenly fluid substance"

[36]Bianchini 1996, p. 158. Letter of 7 September 1726 to Melchior della Briga (1686–1749), a Jesuit mathematician in Florence.

between the observers and the planet:

> This we know has often happened at other times and places in regard to different stars. For this effect it is sufficient that parts of different density should be mixed, as is obvious in the surf becoming whitish, even though it be composed of clear water and transparent air. At those times when the substance has again become rarefied or separated out of the mixture, that whiteness or phenomenon of another colour vanished.[37]

Nor did Jacques Cassini, the son of Cassini senior and his follower as director of the Paris observatory, believe in the moon, such as he made clear in a paper of 1732. Referring to his father's observations of 1672 and 1686, he argued that "it is not likely that this phenomenon is in the atmosphere of this planet since it would be difficult to believe that it would raise to such a height and even less that it would be a satellite because one would not have avoided noticing it since then."[38] As an alternative explanation, Cassini junior suggested what was essentially the same idea as Bianchini's: What had been observed was a temporary condensation of the "celestial fluid matter" supposed to fill the space between the planets. He thought that the interplanetary condensation might have reflected the light and thus appeared as a spot on Venus or a satellite associated with it. It is quite possible that he got this idea from Bianchini's book. Some thirty years later a different version of the hypothesis of celestial fluids was discussed more elaborately by the French physicist Mairan (see section 4.1).

2.3 The satellites of Mars

Within the tradition of natural theology, so popular in the eighteenth century, the satellites or secondary planets were often seen as created by God with a definite purpose. Because Jupiter and Saturn are so far away from the Sun, "so to make them amends, they are surrounded with a more grand Retinue of Secondary Planets, or Moons," wrote the English vicar and natural philosopher William Derham in his widely read *Astro-Theology*. The result was "a very noble, and entertaining scene of the Creator's Glory." He wrote about Saturn's five attendants and suggested that "there is great reason to conclude there is a Sixth lying between the two Outermost, there being a larger space between them than is in proportion to

---

[37] Ibid., p. 159.

[38] Cassini 1735, p. 211. This was the paper in which Jacques Cassini (1677–1756) suggested that the period of rotation of Venus was 23 hours 20 minutes.

what is found amongst the rest."[39]

According to Derham, there was no corresponding need for moons around the interior planets: "As for Venus and Mercury there may be no occasion for any Attendants, by reason of their proximity to the Sun."[40] On the other hand, he thought that Mars might well be provided with one or more small moons:

> [In the solar system] we shall find a no less admirable Scene of the great Creator's Care and Wisdom, than we discovered in the Earth and Moon. In Mars indeed, we can discern a great similitude with the Earth, in its Opacity and Spots, but we have not yet been able to perceive any attendance of Moons, as in the other superior Planets; not so much probably because there are none, but because they are small, or they reflect a weak light, and are at a great distance from us.[41]

The kind of reasoning exhibited by Derham could be found in the more popular literature throughout the century, indeed extending well into the next century. It was a view fully accepted by Johann Bode in his *Betrachtung der Gestirne und des Weltgebäudes*, a classic of pluralist literature published in 1816. God had created the moons to bring light to the distant planets, so their inhabitants could "praise and admire the power and goodness of their eternal Creator."[42] As late as 1854 the physicist David Brewster described the satellites as "domestic lamps which light the primary planets in the absence of the sun."[43] We are not aware of any eighteenth-century author who used natural theology to argue either for or against a satellite of Venus, although some came close. Mars was another matter.

Derham's suggestion that Mars might have a moon was neither the first nor the last of its kind. The first astronomer who claimed to have seen moons around Mars was probably Johann Burchard Schyrle, a Capuchin friar and professor of theology known to his contemporaries as Anton Maria Schyrlaeus de Rheita. In two observations, made on 29 December 1642 and 4 January 1643, he observed what he took to be five new Jovian moons, to which he gave the name *Urbanoctavianes*

---

[39] Derham 1721 (first published 1714), p. 203. William Derham (1657–1735) was elected a fellow of the Royal Society in 1703. Apart from his works on natural theology, he published works in natural history and physics, including an early determination of the velocity of sound.

[40] Derham 1721, p. 194. This was a standard view in the period, shared by Bernard Fontenelle (1657–1757), for example. In his immensely popular *Entretiens sur la pluralité des mondes*, originally published 1686, he stated that the inner planets had no need of a moon; he also expressed his surprise that Mars, although farther away from the Sun than the Earth, did not have its own satellite. Fontenelle 1955, p. 125, a reprint of the 1742 edition. Fontenelle did not refer to a possible Venus moon. Also he did not refer to Cassini's observations of 1672 and 1686 in his extensive *Éloge* of Cassini (Fontenelle 1740, vol. 1, pp. 270–312).

[41] Derham 1721, pp. 193–194.

[42] Bode 1816, pp. 364–365. For Bode (1747–1826) as a pluralist, see Crowe 1999, pp. 73–77.

[43] Brewster 1854, p. 89.

after Pope Urban VIII. He also claimed to have found new satellites orbiting the other superior planets, Mars and Saturn. Rheita reported his startling observations in a letter of 6 January 1643 to Erycus Puteanus, a professor in Louvain, and soon the news was disseminated to other natural philosophers.[44] The published version of the discovery claims appeared later the same year in a book published in Louvain with the title *Novem stellae circa Jovem, circa Saturnum sex, circa Martem non-nullae* (that is, "Nine stars around Jupiter, six around Saturn, and several around Mars"). He repeated the claims in an important work published in Antwerp two years later, *Oculus Enoch et Eliae, sive radius sidereomysticus.* In this book he showed a drawing of Mars surrounded by six flattened moons and also included a map of the Moon. Rheita had become acquainted with Johannes Wiesel in Augsburg in 1643–44, and in *Oculus Enoch et Eliae* he praised the telescopes made by the skilful optician.

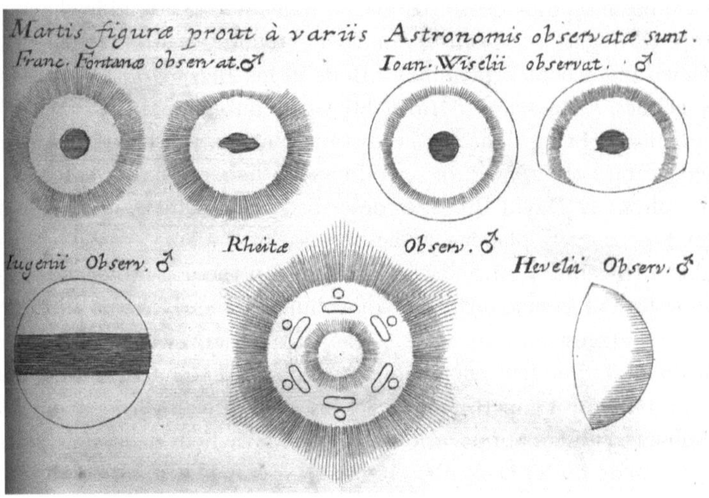

Figure 2.8: Drawings of Mars, as observed by Fontana, Wiesel, Huygens, Rheita and Hevelius, as reproduced in Zahn's *Specula* of 1696. Note Rheita's six moon-like objects around Mars. A similar figure appeared in Rheita's *Oculus Enoch et Eliae* from 1645.

Rheita was for a period well known as an astronomer, but his discoveries of new satellites failed to win recognition.[45] Gassendi argued in a small tract of 1643

[44]Malcolm and Stedall 2005, p. 355. The suggestion that Jupiter was attended by more than four moons did not originate in either Rheita or Fontana. It had been made as early as 1612, by Scheiner in his *Maculis solaribus* (see Shea 1970).

[45]On Rheita (1604–1660), see van Helden 1994 and Willach 2001. It is believed that he went

that Rheita's "satellites" around Jupiter were really fixed stars, an argument which was substantiated by Hevelius in his *Selenographia* of 1647. According to Gassendi and Hevelius, the satellites that Rheita had claimed to discover were really fixed stars in the constellation of Aquarius. By mapping the region of Aquarius and superimposing Rheita's diagram on it, Hevelius showed that the five new satellites were still in the same position, but with no connection to Jupiter. This amounted to strong visual evidence against Rheita's claim.[46]

Before Hevelius' rejection, Rheita defended his claim by pointing out in *Oculus Enoch et Eliae* that the mutual distances of the observed objects changed, so that they could not be fixed stars. Rheita's claim received support from Caramuel Lobkowitz, a Cistercian natural philosopher, whose contribution to the debate was included in the book of 1643. On the other hand, Riccioli did not find the evidence for new satellites to be convincing. In his *Almagestum novum* he reviewed the claims of Fontana and Rheita, and also mentioned that additional moons around Jupiter had been seen by Johannes Baptista Zupus (or Zupo), who used a telescope of Fontana's construction.[47] A Jesuit mathematician and a friend of Fontana, Zupus may have been the first to observe phases of Mercury. According to Fontana, who published his observation, he first saw the phases of Mercury on 23 May 1639.[48]

Apart from the early ideas of Kepler, Rheita, Fontana and Huygens, in the first half of the eighteenth century several other authors mentioned the possibility that Mars was supplied with one or more Moons. Most of these suggestions relied on the physico-theological argument that the inhabitants of Mars were in need of moons to serve them. Thus, in 1711 a German clergyman named Andreas Ehrenberg (or Hareneus Geierbrand) speculated that Mars was endowed with one or two moons, and he was followed a decade later by another German, Johann Jacob Schudt, a school rector.[49]

The influential German philosopher and systematizer of knowledge, Christian Wolff from Halle, wrote a series of *Vernünfftige Gedancken* on a variety of learned topics. In one of these works, published 1723 and dealing with natural objects,

---

into a kind of partnership with Johannes Wiesel, who manufactured some of his telescopes. A crater on the Moon is named after Rheita. Accused by the Inquisition, he was imprisoned in Bologna and later in Ravenna, where he died.

[46]Hevelius 1967 (a facsimile reprint of the original), pp. 33–35. Winkler and van Helden 1993, p. 106.

[47]Riccioli 1651, vol. 1, pp. 488–490. On the suggestions of more than four Jupiter moons, see also Débarbat and Wilson 1989, p. 148.

[48]Fontana 2001, p. 90. Zupus (1589–1667) taught mathematics in the Jesuit college in Naples for 27 years. His name is sometimes given as Zupo or Zupi.

[49]According to Crowe 1999, p. 34. The titles of the works were Ehrenberg, *Curiöse und wohlgegründete Gedancken von mehr als einer bewohnten Welt* (Jena, 1711) and Schudt, *De probabili mundorum pluralitate* (Frankfurt, 1721).

he discussed the distances of the planets from the Sun and also their number of satellites. Like Derham, to whom he referred, he subscribed to the view that the more far away from the Sun a planet was, the more satellites would it have. Also like Derham, he saw no reason why Venus should be equipped with a moon. But matters were different with Mars, an exterior planet: "It may be assumed that Mars possesses one or perhaps more moons, but that they are so small that we cannot discover them from the Earth, not even with our telescopes."[50] In addition, Wolff came up with another proposal for why the supposed Martian moon had remained unobserved, namely that it reflected the light of the Sun very poorly.

As is well known, a few years later Jonathan Swift described in *Gulliver's Travels*, published 1726, how the scientists of Laputa had discovered two satellites revolving around Mars.[51] The relevant passage is this: "They [the Laputians] have likewise discovered two lesser stars, or satellites, which revolve about Mars, whereof the innermost is distant from the centre of the primary planet exactly three of the diameters, and the outermost five; the former revolves in the space of ten hours, and the latter in twenty-one and a half."[52]

Also Voltaire, in his *Micromégas* of 1752, wrote of two Mars moons, undoubtedly inspired by Swift whom he had met in England during his stay 1726–29. In this satirical work, a giant inhabitant – his height is eight leagues or about 35 km – of one of the planets revolving round Sirius embarks on a cosmic voyage, helped by his "marvelous knowledge of the laws of gravitation, and of the forces of repulsion and attraction." He traverses easily from globe to globe, and also happens to visit the tiny Earth. Voltaire used the story to criticize science popularizers such as Fontenelle and also to ridicule Derham and his astro-theology. In company with a much smaller inhabitant of Saturn, Micromégas visits Jupiter and subsequently

> ... they travel through a space of about one hundred millions leagues, and they pass along the planet Mars, which as you know is five times as small as our little globe; they see two moons which serve this planet and which have escaped the eyes of our astronomers. I know very well that father Castel has written, and that quite plainly, against the existence of these two moons; but I relate myself to those who argue by means of analogy. These good philosophers know very well how difficult it would be if Mars, being so far away from the Sun, happened to have less than two moons.[53]

[50] Wolff 1752, §85, p. 142. Originally published 1723.

[51] As mentioned, Kepler suggested two satellites of Mars as early as 1610. On Swift (1667–1745) and the moons of Mars, see Gould 1945, Gingerich 1970 and the updated version Gingerich 1978.

[52] Swift 1994, p. 185.

[53] Voltaire 2000, pp. 41–42. On *Micromégas* and its sources, see Nicolson 1948, pp. 214–219. Swift's *Gulliver's Travels* was translated into French in 1727, becoming an instant success.

It is much less known that in between Swift and Voltaire an obscure German astronomer not only suggested the existence of a satellite of Mars, but actually claimed to have discovered it. To the extent that Eberhard Christian Kindermann, court astronomer to the king of Poland, is known at all, it is probably as Germany's first science fiction writer and not as an astronomer. In 1744 he published a novel, *Die geschwinde Reise auf dem Luft-Schiff nach der obern Welt*, in which he described how five German mariners set off on a journey to explore the solar system. Among the purposes of their planetary voyage was "to find out whether it is true that on July 10 of this year, the planet Mars appeared with a satellite or moon for the first time since the world has been in existence." The preface offered a few more details, including that he had seen the moon between 3 and 4 o'clock in the morning: "Herr Kindermann has discovered it for the very first time, at the said hour of the day and with a tube [telescope] of his own making."[54] However, he only saw it traversing a part of its orbit, before it moved behind the planet.

Two years later Kindermann wrote of a comet which he claimed to have observed, although it is uncertain if it ever really existed. In this small pamphlet he included a figure professing to show the orbit of the satellite of Mars that he had discovered on 10 July 1744. Now the period of revolution was stated to be 59 hours 50 minutes 6 seconds, and the satellite to be removed about 2.5 Mars radii from the centre of the planet. According to the German astronomer, both Mars and its satellites were provided with atmospheres.[55] Of course, this is not where the story of the moons of Mars ends. The long phase of speculation and hypothesis ended in 1877, when Asaph Hall discovered that the planet carries with it two small moons. We shall return to Hall's discovery in Chapter 6.

2.4 James Short and his Contemporaries

We know of no more observations of the elusive Venus moon until 1740, when the young Scotsman James Short turned his reflecting telescope towards Venus. After

Voltaire was familiar with the work, to which he referred in *Micromégas* (p. 52). Louis Bertrand Castel (1688–1757), a Jesuit natural philosopher of the Cartesian school, attacked in several works the Newtonian science. He was known for his invention of an ocular harpsichord, an imagined instrument which employed the analogy between light and sound (see Hankins 1993). In a letter of 1738, Voltaire called him "the Don Quichotte of mathematics." For a later eighteenth-century advocate of Martian moons, see Helmuth 1794, pp. 285–288, as mentioned in section 3.4.

[54]Kindermann 2006, pp. 39–40. The German astronomer was aware of Wolff's prediction, which he quoted with approval.

[55]Kindermann 1746. Curiously, the claim of a Mars moon was not mentioned in the text at all, but only appeared from the figure. Copeland 1892 may have been the first to call attention to the little known Kindermann and his satellite of Mars.

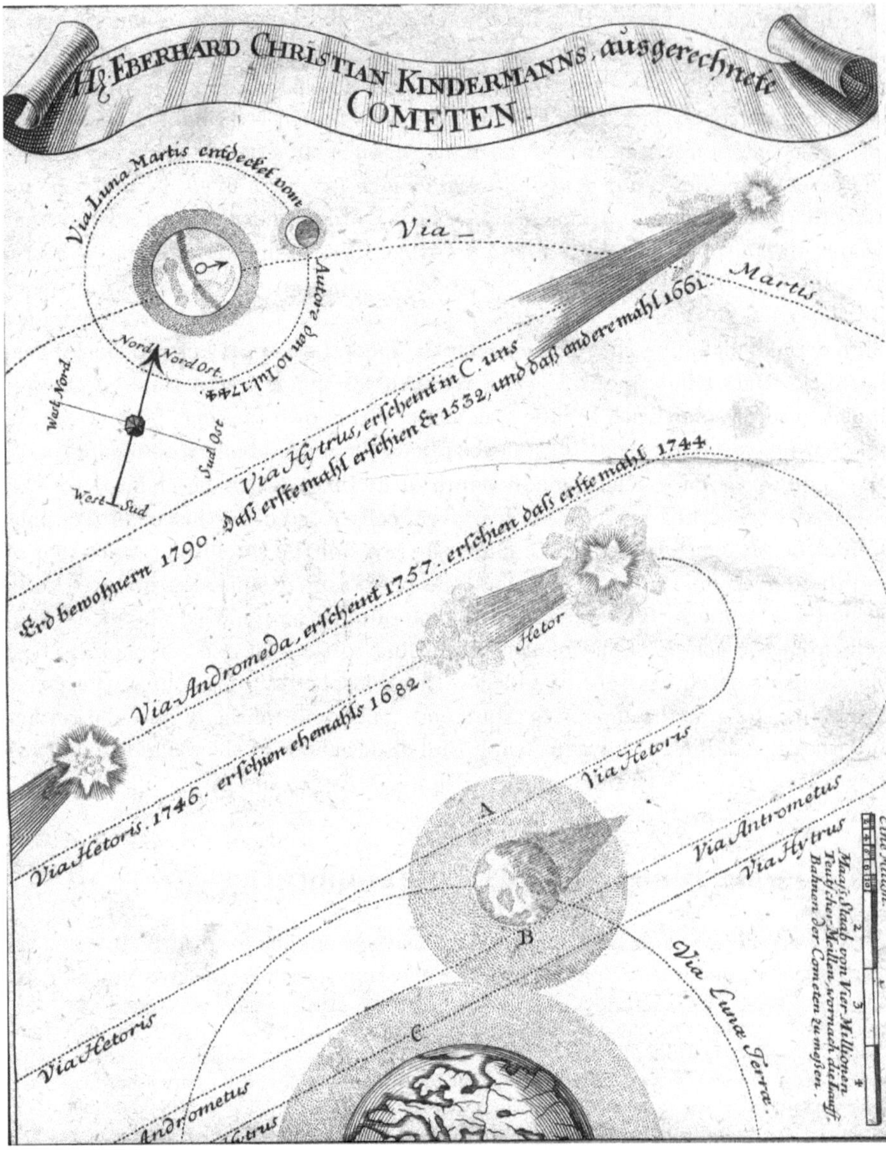

Figure 2.9: The German astronomer Eberhard Kindermann's depiction of the satellite of Mars, such as it appeared in a pamphlet on comets of 1746.

having been elected a fellow of the Royal Society in 1737, he settled the following year in London, where he established himself as a telescope manufacturer of the first rank, specializing in Newtonian reflectors.[56] He soon won recognition not only as an expert manufacturer of reflectors but also as an optician. He was one of the first British natural philosophers to endorse the vibration theory of light (as an alternative to the projectile theory) and to incorporate John Dollond's achromatic lenses into a telescope.

Short published many astronomical observations in the *Philosophical Transactions*, one of the earliest being of "a small Star" very near Venus. With a telescope fitted with a micrometer, he found on the morning of 3 November 1740 (Gregorian style) an object at a distance of about 10 degrees and 2 minutes to the west of Venus:

> Finding Venus very distinct, and consequently the Air very clear, I put on a magnifying Power of 240 times, and, to my great surprise, found this Star put on the same Phasis with Venus. I tried another magnifying Power of 140 times, and even then found the Star under the same Phasis. Its Diameter seemed about a Third, or somewhat less, of the Diameter of Venus; its Light was not so bright and vivid, but exceeding sharp and well defined. A Line, passing through the Centre of Venus and it, made an Angle with the Equator of about 18 or 20 Degrees.[57]

Short studied the object for about an hour, until the sunlight became too strong for further observation. During the following mornings he continued to look for it, "but never had the good fortune to see it again." He was aware of Cassini's earlier observations to which he briefly referred. Short also reported that he had seen "two darkish Spots" on the surface of Venus, but did not give further information. In his brief communication he cautiously avoided referring to the star-like object as a moon or satellite of Venus (except in his title). It is unknown if he originally believed the object to be such a moon, but when he made observations in connection with the Venus transit in 1761 he had nothing to say of either a star, a spot or a satellite.[58] In fact, during a meeting of the Royal Society in 1761, where the recent French observation claims of the satellite were reported, he supposedly pointed out that he did not believe in the satellite of Venus. According to another participant in the meeting, he wanted to retract his earlier paper.[59]

[56]On Short and his career in astronomy and telescope-making, see Turner 1969.

[57]Short 1744, p. 646. England only changed to the Gregorian calendar in 1752, and so Short reported his observation according to the Julian system, as 23 October. See also Anon. 1876a. Short's paper was presented to the Royal Society in January 1741, but not published until 1744.

[58]Short 1762.

[59]See Hell 1792, p. 117, who refers to a conversation he had in 1764 with Nathanael Matthaeus

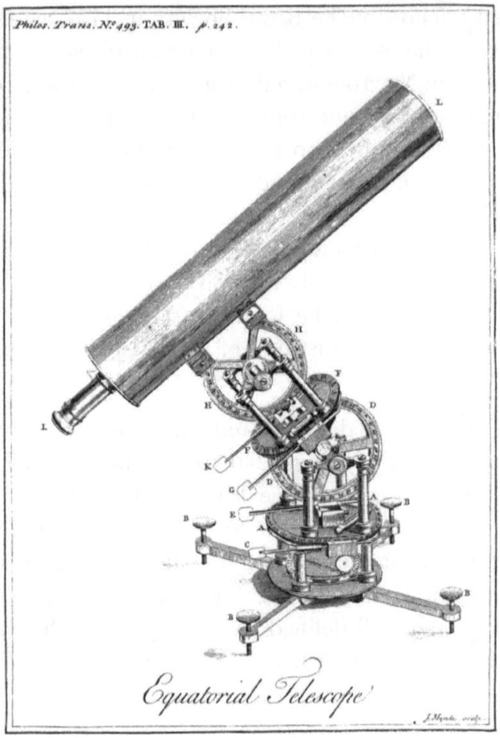

Figure 2.10: One of James Short's popular reflecting telescopes, as described in J. Short, "The description and uses of an equatorial telescope," *Philosophical Transactions* **46** (1749–50), 241–246. On Short and his instruments, see Turner 1969.

The distinguished Paris astronomer Joseph Jérôme Lalande, professor of astronomy at the Collège Royal and since 1753 a member of the Académie des Sciences, was known to have an interest in spectacular phenomena, whether astronomical or not. A man of boundless energy and an insatiable thirst for fame, he had a weakness for the unusual. The interest included the satellite of Venus, about which he wrote at several occasions, in the *Encyclopédie* as well as in his important textbook *Traité d'astronomie.* In the spring of 1763 he made a voyage to London, arriving on 15 March, and already the next day he visited Short. During the stay, he became involved in the Harrison affair, that is, the delayed award of the British government to the instrument maker John Harrison for his invention of

von Wolf (1724–1784), a Polish-German astronomer and foreign member of the Royal Society.

an accurate maritime chronometer.[60] During one of his conversations with Short the Scottish telescope maker confided to him that he no longer thought to have seen a moon. According to Lalande's account:

> Short ..., so it seemed to me, no longer believes in the existence of a Venus satellite but rather in one of some other planet which, reflecting less light, can be seen only with difficulty and rarely. I am convinced that he only made this last supposition in order not to completely abandon a premature opinion which he too expressly had announced in his youth.[61]

On the other hand, some years later Lalande wrote in the *Encyclopédie* a detailed account of Venus' satellite in which he described Short's observation as follows: "This observation, being one of those which best establishes the existence of a Venus satellite by the impossibility of supposing that the observer was deceived by optical illusions, deserves particular attention."[62] Lalande also wrote on the Venus moon in the 206-volume *Encyclopédie méthodique par ordre des matiéres*, a mammoth book project that ran between 1781 and 1832. The three volumes on mathematics and astronomy, with Lalande as the sole author of astronomical subjects, were published 1784–89.[63]

The report of Short, known as an expert observer and designer of telescopes, was well known in continental Europe. Johann Ludwig Oeder, a professor of mathematics and physics at the Collegium Carolinum in Braunschweig, was interested in mirror telescopes of the best possible quality. In a letter of 15 December 1746 to a local optician by the name of Ehrhardt, he wrote that the planned telescope should be able to "make visible the satellite of Venus which a couple of years ago was discovered in England."[64]

Of greater interest is an article in the *Histoire de l'Académie Royale des Sciences* written by the esteemed French physicist Jean Jacques d'Ortous de Mairan, who served as sécretaire perpétuel of the Académie des Sciences in 1741–43 and from 1746 to his death in 1771 was pensionnaire géomètre.[65] Mairan had been

[60]On Lalande's journey to London and his meetings with Short and Harrison (1693–1776), see Chapin 1978.

[61]Lalande 1792, vol. 3, p. 210. This was the third edition of *Traité d'astronomie* (a second edition appeared in 1771).

[62]Encyclopédie II, vol. 35, p. 257. Contrary to the original edition, the Berne-Lausanne edition contained a separate and detailed entry on the moon of Venus (pp. 256–260), including most of the observations of 1761 and 1764. On the later editions of the *Encyclopédie*, see Darnton 1979.

[63]On the *Encyclopédie méthodique* and its troubled history, see Watts 1958.

[64]Quoted in John 2004, p. 61.

[65]Mairan 1744. For its authorship, see Mairan 1764, p. 162. Contrary to the articles in the *Mémoires*, those in the *Histoire* always appeared anonymously, but they were all written by the sécrétaire perpetual. For Mairan it was a burden to write not only the summaries of the major memoires but also the eulogies of deceased members of the Academy. For more details

informed by a letter that Short wrote on 8 June 1741 to Pierre Coste, a French writer staying in England, and in which Short admitted his lack of success in further observations of the supposed moon (at the time when Mairan wrote his report, the issue of *Philosophical Transactions* had not yet appeared in print).[66] Mairan referred to the works of Cassini and Short, but was unable to provide information that could throw new light on the question. "It is surprising," he wrote, "that [although] M. Cassini has made many researches since then at different times to establish a discovery of such a great importance, he was never successful, and for a period of 54 years, no other astronomers, as far as we know, could see the phenomenon, not even M. Bianchini, so renowned for his discoveries concerning the planet Venus."[67]

In the absence of new observations, Mairan reflected at length on the probability of a Venus moon as seen from a philosophical and physical perspective. Like other contemporary authors, he reasoned from analogies and final causes which told him that the satellites must fulfill a purpose for the inhabitants supposed to live on the planets. However, Mairan did not regard analogies of this kind to be quite satisfactory. The moons around the distant planets Jupiter and Saturn served as mighty lamps, but why should nature, or God, have chosen to provide Venus with a moon? And why not Mars? In the *Histoire* for 1741 Mairan referred critically to the received view that satellites served as supplement to sunlight and, therefore, the number of satellites increased with the distance of the planet from the Sun. Not only was this view based on "the convenience of analogies, and the inclination we have to let nature act according to our views and our needs," it also disagreed with the fact that apparently Mars had no satellite. Although he did not doubt that Venus and Mars were inhabited, he was reluctant to draw astronomical conclusions from the supposed needs of their inhabitants.

Most astronomers came to accept Mairan's argument, and in the nineteenth century the older view was often ridiculed and presented as an example of anthropomorphic error.[68] However, the view did not die easily and in connection with

about Mairan as secretary and scientist, see Kleinbaum 1973 and the *éloge* in Fouchy 1774. The suggestion in *Nature* of 1876 that the author was Jacques Cassini, the son of Jean Dominique Cassini, is erroneous (Anon. 1876a).

[66]Pierre Coste (1668-1747) was a French Protestant refugee who lived in England as a writer and translator. He made in 1700 a French translation of John Locke's *Essay Concerning Human Understanding* and later of other of Locke's works. An important ambassador of English thought abroad, he also translated Newton's *Opticks* into French. The French edition, entitled *Traité d'optique*, was published 1720 in Amsterdam.

[67]Mairan 1744, p. 125.

[68]For example, Frederik Kaiser (1802-1878), director of the Leiden Observatory, scornfully referred to "the silly explanation that the more distant planets are in need of many satellites in order to compensate for the sunlight." He pointed out that such an explanation raised the question of why Mars had no moon. Kaiser 1867, p. 167, a Danish translation by Mathilde

arguments of natural theology it can be found well into the nineteenth century. An example will be given in section 5.2.

Concerning the interior planets, Mairan wrote that "it was believed that they did not have any [satellites], and that they ought not to have any, being much closer to the Sun than the Earth; they are also very often confounded in their rays, and this fact does not increase insignificantly the difficulty to discover that which surrounds them." Mairan also had his saying about Venus' satellite allegedly discovered by Cassini:

> A celestial body so difficult to discover from the Earth does not seem to be made for us, and one would not know how to defend a conclusion other than it is intended to enlighten another world and other inhabitants. In this respect the analogy in general leaves nothing to be desired. But let us here point out that this analogy ... was refuted by the planet Mars, more remote than we, and yet it has no satellite, and it is furthermore now refuted by the planet Venus, less remote than we and which has one [satellite] just as big as our Moon.[69]

Although Mairan did not come up with a definite answer, he was favourably inclined towards the existence of a Venus moon, such as first given observational support by Cassini and subsequently by Short (he ignored Fontana). At the end of his essay he referred to the generally accepted view that the interior planets were enveloped by an extended solar atmosphere. If this solar substance was luminous, the darkness at night of these planets would be much reduced. The supposed satellites would probably also be surrounded by this substance, which would change their density and extension as seen from the Earth and thus cause many uncertainties and errors in the observations of the satellites. Mairan would return to the subject in a more elaborate manner twenty years later (see section 4.1).

Proceeding chronologically, the next astronomer to observe the moon of Venus was Andreas Mayer in 1759, although his observations escaped public attention until 1778 and even then lay practically dormant until they were brought forward again in a book of 1875. Mayer, who since 1741 had served as professor of mathematics, physics and astronomy at the University of Greifswald in Northern Germany, then under Swedish rule, reported in his observation diary as follows for 20 May 1759:

Ørsted, the daughter of H. C. Ørsted, of the third edition of Kaiser's *De Sterrenhemel* (Leiden, 1860).

[69] Mairan 1744. A German translation appeared in the *Hamburgisches Magazin* (Mairan 1747–48).

> In the evening, about 8h 45′ 50″, I saw above Venus a little globe of far inferior brightness, about $1\frac{1}{2}$ diam. of Venus from herself. Future observations will show whether this little globe was an optical appearance or the satellite of Venus. The observation was made with a Gregorian telescope of thirty inches focus. It continued for half an hour, and the position of the little globe with regard to Venus remained the same, although the direction of the telescope had been changed.[70]

At the time of the Venus transits of 1761 and 1769, Mayer continued his observations, perhaps in the hope of establishing whether or not the "little globe" was a satellite of Venus. However, one may surmise that he did not see the object again, for in his publication reports of the two transit observations he did not mention any satellite.[71] On the other hand, in his report of the first event he included a list of Venus observations made between November 1758 and February 1762, and in a footnote he did refer to the "little globe" seen on 20 May 1759. He said as follows: "When on the 20th in the evening I ... directed it [the telescope] towards the evening star, then so lively sparkling, there appeared in that part of the sky at a distance of $1\frac{1}{2}$ diameter a little sphere whose diameter was equal to $\frac{1}{4}$ of that of Venus. ... Whether or not this satellite belongs to Venus, I do not dare claiming."[72]

The observations of Cassini and Short were briefly discussed in the first edition of the famous *Encyclopédie* or, to give its full title, *Encyclopédie ou dictionnaire raisonné des sciences, des arts et des métiers*. The author of the article on Venus, none other than Jean le Rond d'Alembert, maintained an uncommitted attitude: "If it is a Venus satellite, it becomes even more difficult to tell what can be the usefulness of satellites. Would it be so to speak to supplement the light that the planets do not receive from the Sun? But here is a planet much closer to the Sun than we, and it has one [satellite] greater than ours; besides Mars does not seem to have a satellite, although it is farther away from the Sun than the Earth."[73]

---

[70] Quoted in Lambert 1776, p. 186. See also Schorr 1875, p. 65. Moore 1956, p. 93, mistakes Andreas Mayer (1716–1782) for the better known Tobias Mayer (1723–1762), the astronomer and cartographer at Göttingen University. (The error does not appear in Hunt and Moore 1982.) Nor is Andreas Mayer to be mixed up with the contemporary astronomer Christian Mayer (1719–1783), who did most of his work in Heidelberg. Greifswald in Pomerania came under Swedish rule in 1648, as a result of the Thirty Years War. In 1815 the city became part of Prussia.

[71] Mayer 1762; Mayer 1769. The observations of Venus' emersion on 5 June 1761, reported in Mayer 1762, were made by his collaborator Lampert Heinrich Röhl (1724–1790) who in 1773 became director of the Greifswald Observatory.

[72] Mayer 1762, pp. 16–17.

[73] Encyclopédie I, vol. 17, p. 34. It appears from the article on Venus that it dates from July 1760, that is, before the transit. Most of the articles in the *Encyclopédie* were marked by a letter by means of which the author could be identified (see the list in volume 1, p. lxxxix). For

It is worth pointing out that in the early observations of and comments on the satellite of Venus, Fontana's original observations were nearly invisible. Cassini did not mention the Neapolitan astronomer, and neither did Short, Mairan nor Gregory (Bianchini was the only exception). It is hard to believe that Cassini and Gregory were unaware of Fontana's observations, but they may have decided that they were not reliable.

example, the letter "O" referred to d'Alembert.

Chapter 3

From climax to anticlimax

Until 1760 there had only been eight observations that related to the supposed moon of Venus, and they were scattered over more than a century. The situation changed drastically in the following year, which should undoubtedly be seen in the context of the long-awaited Venus transit of 1761, an event that focused international astronomical attention on the course of the planet.[1] It is possible that transits of Venus have been observed before the invention of the telescope, for under the right circumstances a Venus transit should be visible to the naked eye. When the planet passes in front of the Sun, its apparent diameter is about 1/31 of that of the Sun. However, claims of pre-telescopic observations remain speculative.[2]

The real history of the phenomenon only starts in 1630 with Kepler's prediction based on the new *Tabulae Rudolphinae* that Mercury and Venus would both transit the Sun in 1631, on 7 November and 6 December, respectively. Ever since this prediction, and especially since Edmond Halley (and, before him, Jeremiah Horrocks) had shown that the passage of Venus could be used to measure the distance of the Earth from the Sun, there had been great interest in these rare events. Before 1761, only one transit of Venus had been observed, namely when Horrocks and his friend William Crabtree saw the planet moving across the face of the Sun on the 24th of November 1639 (Julian system).

[1]The literature on Venus transits is considerable. See Woolf 1959 and Sheehan and Westfall 2004, neither of which works refers to the satellite of Venus and its connection to the transit observations. See also the bibliographies avalilable online, such as `http://www.phys.uu.nl/~vgent/venus/venus_menutext.htm`.

[2]See the critical discussion in Goldstein 1969 of possible transit reports made by Avicenna, Averroes and other Islamic astronomers in the Middle Ages. Kepler and some of his contemporaries believed that a Mercury transit had been observed in Europe in the ninth century (van Helden 1976b and Sheehan and Westfall 2004, pp. 57–59).

The 1761 transit results were mainly based on Halley's method, which required measurements of the duration between the contacts observed from two different stations. They were somewhat disappointing, in the sense that they did not succeed in narrowing down the solar parallax very precisely. The results ranged from $8''.3$ to $10''.6$. The larger and better organized 1769 observation programme fared better, as computations based on the many observations gave results in the interval from $8''.50$ (Lalande's value) to $8''.88$ (Pingré's value). The corresponding solar distances were 154.726 million km and 148.108 million km. The current best estimate of the astronomical unit is close to 149.598 million km.

One of the main reasons for the uncertainty of the solar parallax was difficulties in exact timing caused by the so-called "black drop" effect. First publicly described by the Swedish chemist and physicist Torbern Bergman in 1761, it turned out that Venus did not separate cleanly from the edge of the Sun, but that it appeared for a brief while to be connected to the edge with a thick filament. The disc of Venus was found to assume a distorted form, as if it was glued to the Sun's edge. Moreover, since the phenomenon varied from observer to observer, it caused problems with establishing the exact time of the transit. The black drop effect continued to be controversial during the transits in the nineteenth century. It was for a long time believed that the effect was caused by either optical illusion, diffraction of light around Venus, or refraction in the Venusian atmosphere. However, none of these explanations are correct. Only in recent years has the phenomenon of the black drop been fully understood as due to a complex smearing mechanism.[3]

Although the determinations of the solar parallax in 1761 were not entirely satisfactory, as a side effect they resulted in valuable knowledge of the behaviour and nature of planet Venus. Among these side effects were attempts to observe the hypothetical satellite that had previously been seen by Fontana, Cassini, Short and Mayer. Half of all observation claims of the Venus moon, namely 19, occurred in that year, with the most important ones being recorded in Limoges and Copenhagen. Nine more observations followed in 1764–68, and then it was over. In the table below we summarize all the known observations, a total of 36, made in the 123 years between 1645 and 1768.[4]

Astronomers interested in the question of a Venus moon realized that the 1761 transit provided a unique opportunity of either confirming or disconfirming the

[3]On the correct explanation of the black drop effect, see Schaefer 2001. Something like the modern understanding of the effect was first proposed by Lalande in 1770, although his explanation was incomplete (Lalande 1773).

[4]More complete lists, including details of the observations, can be found in Haase 1860–69, p. 8, Stroobant 1887a, pp. 6–9, and Schorr 1875, pp. 66–67. The first list of observations of the Venus moon appears in Lambert 1775, p. 186, and includes data from Cassini in 1672 to Montbarron in 1764.

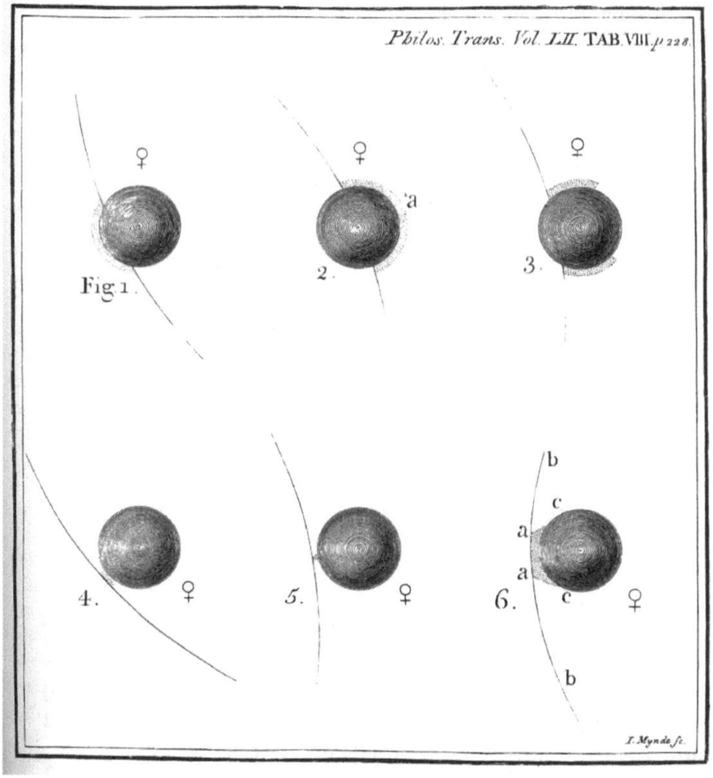

Figure 3.1: The black drop effect as depicted by Torbern Bergman, a famous professor of chemistry at the University of Uppsala, Sweden. Although best known as a chemist within the tradition of the phlogiston theory, Bergman also made important works in physics, geology and astronomy. He recounted his observations of the Venus transit in a letter of 19 November 1761 to Benjamin Wilson, F.R.S., published in *Philosophical Transactions* **52**, part 1 (1761), 227–228.

hypothesis. One of those astronomers was James Ferguson, the Scottish astronomy author and designer of astronomical instruments. In his widely read *Astronomy Explained upon Sir Isaac Newton's Principles*, a work of 1756 that went through seventeen editions, he wrote:

> Venus may have a Satellite or Moon, although it be undiscovered by us; which will not appear very surprising, if we consider how inconveniently we are placed for seeing it. For it's enlightened side can never

Year	**Published**	**Observer**	**Place**	**No. of observations**
1645	1646	F. Fontana	Naples	3
1646	1646	F. Fontana	Naples	1
1672	1730	J. D. Cassini	Paris	1
1686	1730	J. D. Cassini	Paris	1
1740	1744	J. Short	London	1
1759	1762/1778	A. Mayer	Greifswald	1
1761	1781	L. Lagrange	Marseille	3
1761	1761	J. Montaigne	Limoges	4
1761	1761	(anonymous)	St. Neot	1
1761	1813	F. Artzt	Gundersløvholm	1
1761	1776	A. Scheuten	Crefeld	2
1761	1882	P. Roedkiær	Copenhagen	8
1764	1765	P. Roedkiær	Copenhagen	2
1764	1765	C. Horrebow et al.	Copenhagen	3
1764	1766	Montbarron	Auxerre	3
1768	1882	C. Horrebow et al.	Copenhagen	1

Table 3.1: Observations of Venus' moon.

> be fully turned towards us, but when Venus is beyond the Sun; and then, as Venus appears little bigger than an ordinary Star, her Moon may be too small to be perceived at such a distance.[5]

As a believer in the divine principle of plenitude, he could easily imagine that God had chosen to provide the planet with a satellite. After having explained the difficulties of observation, Ferguson continued: "But if she has a Moon, it may certainly be seen with her upon the Sun, in the year 1761, unless its Orbit be considerably inclined to the Ecliptic; for if it should be in conjunction or opposition at that time, we can hardly imagine that it moves so slow as to be hid by Venus all the six hours that she will appear on the Sun's Disc."[6] Alas, no moon materialized. A footnote in the sixth edition of 1778 tersely informed the reader that "The transit is over since this was wrote, and no Satellite was seen with Venus on the Sun's Disc."

[5] Ferguson 1778, p. 18, the sixth edition, corrected and with an appendix on the results of the 1761 Venus transit observations.
[6] Ibid.

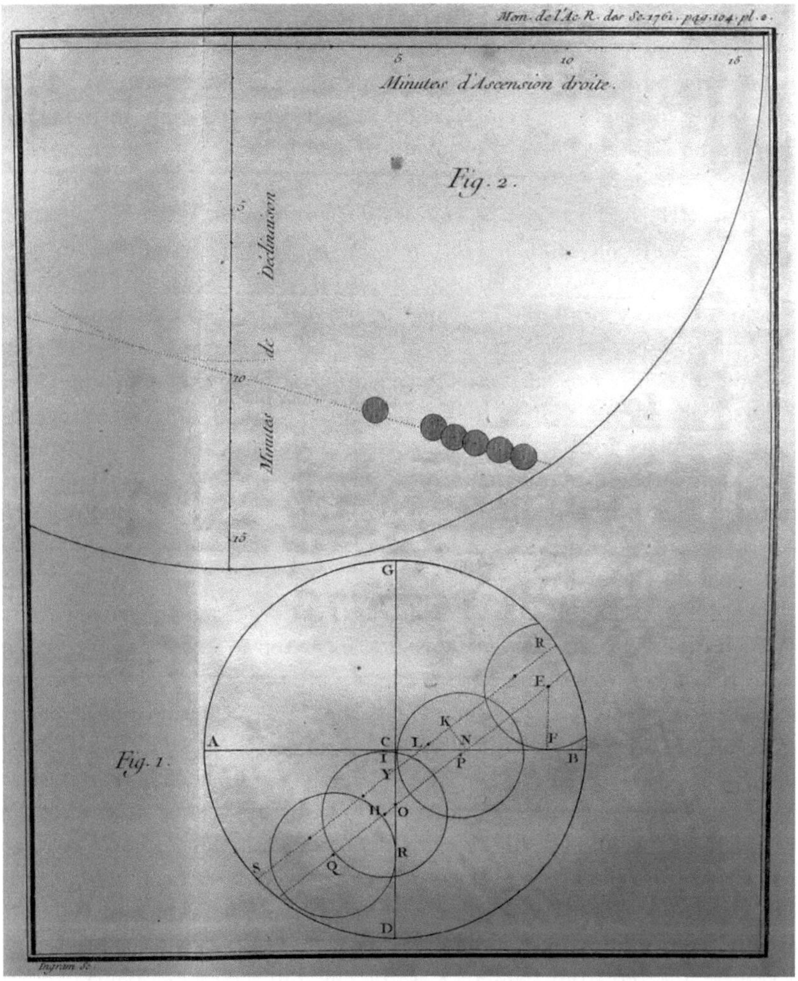

Figure 3.2: One of many illustrations of the Venus transit of 1761, this one shows the series of observations made by de Fouchy and Ferner in the Chateau de la Muette. On Fig. 1 is shown the Sun's passage over the telescope's field of sight, and in Fig. 2 the positions of Venus at six different occasions. From Grandjean de Fouchy, "Observation du passage de Vénus sur le Soleil," *Mémoire de l'Académie Royale des Sciences* **63**, 1761 (published 1763), 96–104.

3.1 The 1761 Venus transit

Louis Lagrange, a French-Italian Jesuit and an accomplished astronomer, had no relation to the famous mathematician and scientist of the same surname, Joseph-Louis Lagrange. Almost nearly forgotten today, he was well known in the latter half of the eighteenth century and the fifth to report observations of the satellite of Venus. He had worked as the assistant of Esprit Pezenas, a professor of hydraulics in Marseille, and in 1759 he met with the Croatian-Italian polymath Roger Boscovich who wanted him to come to Pavia and organize the teaching of Astronomy. Three years later Lagrange was called from the Marseilles Observatory to help with the establishment of the new observatory at the Jesuit college at Brera near Milan. He was entrusted with the scientific direction of the institution, while Boscovich played a more informal, if no less important role. When in 1772 Lagrange was placed in charge of the Brera Observatory and Boscovich relieved from his duties, the disagreements between the two evolved into open enmity.[7] Lagrange left the Brera Observatory in 1776, after having published *Mémoire sur la longitude du College de Brera à Milan*, a history of the observatory in which he did not mention Boscovich at all.

While in Marseille, Lagrange made three observations of Venus' satellite between 10 and 12 February, 1761. He used a 6 feet refracting telescope constructed by Short five years earlier and with a magnifying power of 800. According to Lalande, who apparently was in contact with Lagrange: "He did not see there any phase, as all the other observers had; and it is no less surprising that it seems that this little star followed a path, perpendicular to the ecliptic. This direction, which can be settled from the preceding observations in Limoges, seemed so strange to P. la Grange that he did not find it difficult to abandon all the consequences which he had drawn from these observations."[8] As far as we have been able to establish, Lagrange did not himself communicate his observations. Apparently he did not believe that what he had seen was really a satellite of Venus. This is not only what can be inferred from Lalande's statement, but also what Lagrange wrote in letters to Maximilian Hell in Vienna (see section 4.2).

A few months later another Frenchman, Jacques Montaigne, made observations of the Venus moon, this time in Limoges in central France. His work was

[7] The Jesuit order was disbanded by Pope Clement XIV in 1773. On the strained relationship between Lagrange and Boscovich, see Hill 1961, pp. 81–93. Lagrange published several astronomical and meteorological observations, but apparently not on the satellite of Venus (Poggendorff 1863). Pezenas (1692–1776), his mentor, became director of the observatory in Avignon, where he specialized in solar research and in 1774 published *Nouvelles théorie du Soleil.*

[8] Reported by Lalande in his article on the Venus moon in Encyclopédie II, p. 259. On the Limoges observations, see below. Lalande praised Lagrange as an astronomer recognized for his experience and accuracy, for which reason his observations deserved to be taken seriously.

closely related to Armand Henri Baudouin de Guèmadeuc, a Paris civil servant (maître de requêtes) and a somewhat shadowy figure in French scientific and cultural life. Born in Colmar on 17 April 1737, Baudouin lived as a young man with his uncle, who served as a canon at the Notre Dame in Paris. He subsequently had a varied career in the magistracy, in literature and in astronomy. Charged with theft, in 1779 he lost his job and was imprisoned for fifteen months, first in Vincennes and then in Tanlay. Later he was forced to leave France, apparently after having been involved in a minor episode of bribery. In 1782 Baudouin published anonymously in Neuchâtel *L'Espion dévalisé*, a collection of scandalous anecdotes that caused considerable furor and was reprinted in 1783 and 1784. For a long time this work was ascribed to Honoré Mirabeau, the French count and politician who played such an important role in the early phase of the revolution. Baudouin de Guèmadeuc died in Paris in 1817.[9]

Baudouin seems to have known several of the period's men of science and letters. One of them was Jean-Hyacinthe Magellan, a Portuguese expert in scientific instruments, with whom he corresponded in the 1780s.[10] He was also an acquaintant of the Italian diplomat Ferdinand Galiani, who from 1759 to 1769 served as secretary of the Neapolitan embassy in Paris and eagerly took part in the city's social and cultural life. After Galiani returned to Naples, he wrote to Baudouin, urging him to respond and tell about his life.[11] The letter is not concerned with science, except for a brief passage: "How is the astronomy doing? Has nothing been rescued from the unfortunate voyage of Abbé Chappe? I think he was poisoned together with his fellow travellers." This was a reference to the French astronomer Jean-Baptiste Chappe d'Auteroche, who in 1769 directed a French-Spanish scientific expedition across Mexico to the Californian peninsula. Although the team arrived in time at San José del Cabo and succeeded to measure the Venus transit, the expedition ended tragically: Except for two, the artist Alexander-Jean Noël and an engineer by the name Pauly, they all died from an epidemic.[12] Contrary

[9]Not much is known about Baudouin de Guèmadeuc (1737–1817), but see the account in Michaud 1857, p. 37, where his year of birth is given as 1734. According to George Sarton, he was a benefactor of the pioneer historian of mathematics Jean-Etienne Montucla (1725–1799). When Baudouin was calumniated and exiled, Montucla defended him (Sarton 1936, p. 527) and in 1802 he took care to include him in his *Histoire des mathématiques* (vol. 4, p. 16). See also the editorial note in Galiani 1818, pp. 53–54.

[10]The library of the American Philosophical Society owns three letters from Magellan (1722–1790) to Baudouin dating from 1783–85 (`http://www.amphilsoc.org/library/mole/m/magellan.xml`).

[11]Galiani to Baudouin, 20 April 1770, in Galiani 1818, vol. 1, pp. 53–59. Most of the letters are addressed to Louise-Florence d'Épinays (1726–1783), the French writer and close acquaintant of Jean-Jacques Rousseau (1712–1778) and the circle around the *Encyclopédie*.

[12]Eight years earlier, Chappe (1722–1769) had made an equally adventurous expedition to Siberia, where he made transit observations in Tobolsk. On his involvement in the Venus transit

to what Galiani thought, no poisoning was involved.

Greatly interested in astronomy, in the spring of 1761 young Baudouin had looked for Venus' companion with his 25-foot telescope, but with no luck. He consequently addressed Monsieur Epine, the secretary of the Limoges society of science which had been founded two years earlier by Baudouin's uncle, the quartermaster general Pajot de Marcheval. Through Epine, he got in contact with Jacques Leibar Montaigne, a 45-year-old astronomer. He persuaded him to search for Venus' satellite in connection with the passage of the planet over the Sun, predicted to be visible in early June. When asked to look for a satellite of Venus, at first Montaigne expressed reservation as he did not believe in its existence.[13] Montaigne was known as an ardent comet hunter and later, in 1772 and 1774, he made his name by discovering two comets; in 1780 he claimed to have discovered yet another comet, to be mentioned below.[14]

Using a 9-foot telescope with magnification 40–50, but not provided with a micrometer, Montaigne found the satellite in four observations between May 3 and May 11. He was unable to repeat the success during the transit a month later. According to Lalande's account in the *Encyclopédie*: "He thus saw on 3 May at $9\frac{1}{2}$ in the evening about $20'$ from Venus a little, rather faint crescent and situated in the same way as Venus. The diameter of this little crescent was almost a quarter of that of the planet, and the line drawn from the center of Venus to that of the satellite made with the vertical of that planet and below it towards south an angle of about $20°$."[15] The other three observations gave roughly but not entirely comparable results. The last one in particular differed somewhat from the others.

Having received Montaigne's data, Baudouin read in May 1761 two papers on the subject to the Royal Academy in Paris. He also described it in detail in two treatises published at the same time, which he sent to Maximilian Hell in Vienna, among others. Baudouin and Hell subsequently engaged in a correspondence concerning the satellite, but the precise content of their letters is unknown. Knowledge of the Montaigne-Baudouin work was further disseminated by a German translation of 1761, made on the instigation of Leonhard Euler in his capacity

projects, see Woolf 1959, pp. 115–126, 157–159. His posthumous account of the Californian expedition appeared as Chappe d'Auteroche 1772 and six years later as an English translation, *A Voyage to California to Observe the Transit of Venus* (London: Edward & Charles Dilly, 1778).

[13] According to Hell 1792, p. 20.

[14] Even less is known about Montaigne than about Baudouin, except that he, an able observer of comets, was born on 6 September 1716 in Narbonne (Lalande 1803, p. 477). He is listed in Poggendorff 1863, but without a first name and with no publications. See also Lynn 1884 and Thirion 1885, pp. 45–46. On his comet of 1772, today known as 3D/Biela, see section 6.1.

[15] Encyclopédie II, p. 258.

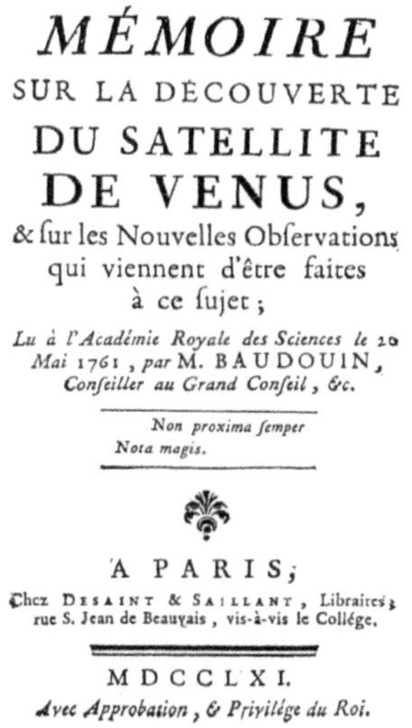

MÉMOIRE
SUR LA DÉCOUVERTE
DU SATELLITE
DE VENUS,
& ſur les Nouvelles Obſervations
qui viennent d'être faites
à ce ſujet;
Lu à l'Académie Royale des Sciences le 20 Mai 1761, par M. BAUDOUIN, *Conſeiller au Grand Conſeil, &c.*

Non proxima ſemper
Nota magis.

A PARIS;
Chez DESAINT & SAILLANT, Libraires, rue S. Jean de Beauvais, vis-à-vis le Collége.

MDCCLXI.
Avec Approbation, & Privilége du Roi.

Figure 3.3: Front page of Baudouin's first memoir on the satellite of Venus, read to the Académie des Sciences on 20 May 1761. The observations on which Baudouin based his claim were solely due to Montaigne.

as director of the mathematical class of the Berlin Academy.[16] It also appeared in an English translation, made by the astronomer John Bevis, a fellow of the Royal Society and a close collaborator of James Short. The translation was published in the short-lived *Mathematical Magazine and Philosophical Repository*, a journal edited by Georges Witchell and Thomas Moss.[17]

We get a little further information from an introductory passage in Hell's *De satellite Veneris* of 1765, in which the Viennese astronomer mentioned the

[16]Baudouin 1761c which includes translations of both of the French memoirs. On Hell's reception of Baudouin's memoir, see Hell 1766, p. 11. Among those who possessed the German text, and probably studied it, was Immanuel Kant, cp. `http://web.uni-marburg.de/kant//webseitn/ka_lek01.htm`. Kant scholars agree that he only acquired books that he actually read.

[17]Only five monthly issues, all of 1761, appeared of the *Mathematical Magazine*. On Bevis (1695–1771), see Wallis 1982. In a booklet of 1883, the amateur astronomer Leeson Prince reproduced Bevis' translation of the Baudouin-Montaigne work. See Prince 1883 and also Lynn 1887a.

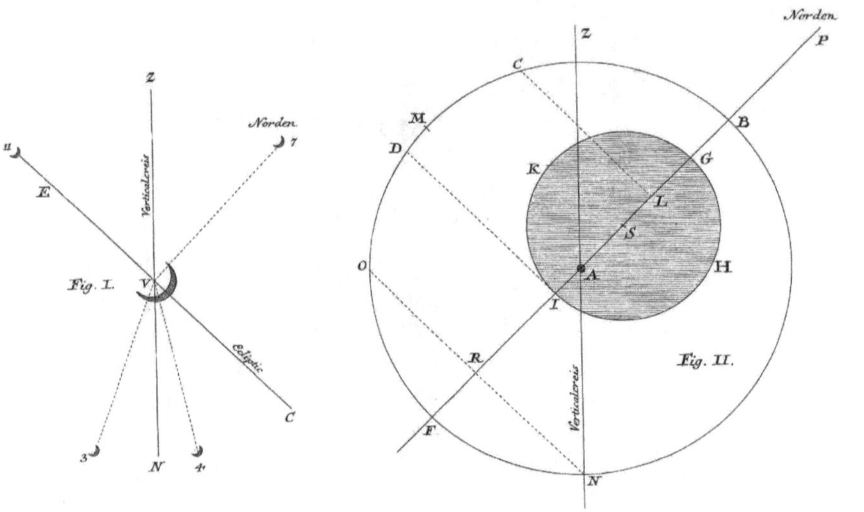

Figure 3.4: The satellite of Venus, as depicted in Baudouin's memoir of 1761. Figure I gives the first three observations of Montaigne. Venus (V) is in the centre and the smaller crescents mark the positions of the satellite on the dates of May 1761 given by the respective numbers. ZN is the vertical and EC a parallel to the ecliptic. The diameter of the satellite was said to be about four times less than the Venus diameter. The distances from V to 3, 4 and 7 were found to be 20 minutes, 21 minutes and 26 minutes, respectively. Assuming the orbit to be circular, Baudouin concluded from these observations that the satellite's synodic period of revolution was about 9 days 7 hours. In figure II, S is the centre of the Sun and *GHIK* the circumference of the solar disc. The dot A marks Venus in front of the Sun, and *BZCDFN* is the orbit of the Venus satellite.

material he was going to include and not include in his work: "Nothing either about the letters sent to me about this matter by the famous M. Wargentin; a few things from the letters of the famous M. Messier, again nothing about the letters of our father La Grange in which he told me that he had never seen this satellite of Venus; nothing about the paper of the famous M. de Mairan on this matter, and nothing about the paper of the renowned M. de Baudouin of Berlin translated into German and annotated by the famous M. Reccard."[18] Hell apparently believed

[18]Hell 1765, p. 6. Wargentin (1717–1783) was a prominent Swedish astronomer, see section 4.2. Gotthilf Christian Reccard (1735–1798) was at the time he translated Baudouin's work a school inspector in Berlin. After having moved to Königsberg in 1765, he was appointed professor

that Baudouin was either a German or associated with the Prussian Academy in Berlin.

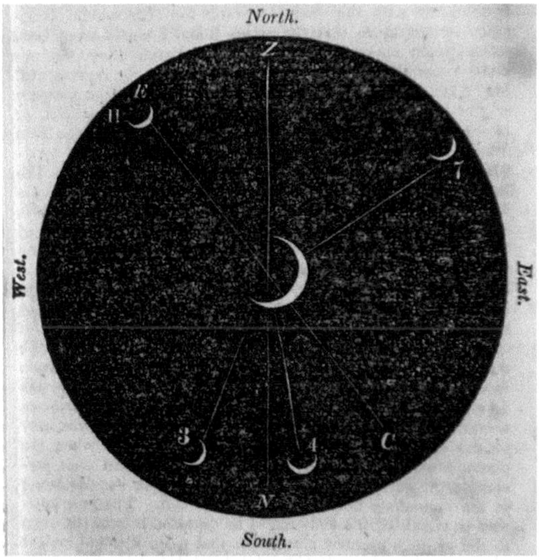

Figure 3.5: Thomas Dick's reconstruction of Montaigne's observations, as given in his *Celestial Scenery* of 1838 and based on Baudouin's memoir of 1761. ZN is a vertical and EC a parallel to the ecliptic. The satellite moving counterclockwise, the numbers 3, 4, 7 and 11 refer to its position and appearance on the respective days. It was first seen on 3 May and the next day it had described an arc of about 30 degrees.

Baudouin was convinced that what he reported was a genuine and important astronomical discovery, and he presented it as if it were the culmination of a long series of attempts to confirm the original discovery of the great Cassini. "Since the year 1686, when Cassini thought to have noticed a satellite near Venus, all astronomers have searched zealously for it," he exaggerated. Only now could it be confidently claimed that "The satellite of Venus is no longer a matter of uncertainty." Although he had not himself seen the moon, he had no doubts about its existence: "It is certain that Venus has a moon and we hope unceasingly to see it."[19]

of theology and thus became a colleague of Kant. Better known as an astronomer than as a theologian, he built his own observatory and published several works in astronomy, including a treatise on the solar eclipse observed on 1 April 1764.

[19]Baudouin 1761c, p. 3 and p. 31. According to Moore 1956, p. 128, Baudouin also reported on

Now France had finally been secured the honour that rightfully belonged to Cassini and to which the Englishman Short (who was actually a Scotsman) had significantly contributed. Alluding to the Seven Years War, in which England and France were engaged as enemies, Baudouin praised eloquently Short for "having raised above the national hatred and made a solemn testimony for France" – one might imagine that Short would rather have been without the honour.

On the basis of the three first observations Baudouin found the period of revolution to be 9 days 7 hours, but after having taken the observation of 11 May into consideration he revised the figure to about 12 days. As to the size and distance of the "new star," he reported that its diameter was one-quarter of that of Venus and that it was almost as far away from Venus as the Moon from the Earth. Confident of the results, he predicted how the moon would move across the disk of the Sun on the day of the transit and even provided an illustration of its orbit. However, when the day of the transit arrived on 6 June, his prediction was not confirmed. On this day Baudouin, in company with Charles Messier, made transit observations in Paris at the terrace of Julian's Baths near the Hôtel de Cluny, the site of Messier's astronomical observatory. He observed the exterior contact between Venus and the Sun, but the satellite eluded him.[20]

In reports of 1768 in the *Mémoires* of the Royal Academy he communicated some of his observations of Venus and also of the moons of Jupiter, without any mention of the elusive satellite of Venus. He compared his own observations with those of Tobias Mayer in Göttingen in order to determine the exact difference in longitude between Paris and Göttingen. For the solar parallax he obtained a best value of 9′30″.[21]

Like several other astronomers, Baudouin was impressed by the similarity between Venus and the Earth, which he now believed had been reinforced.

> It follows that the distance of the satellite from Venus, or half the diameter of its orbit, is almost 60 times the half-diameter of its primary planet, just as the distance of the Moon from the Earth makes up 60 times the half-diameter of the Earth. This provides us with another

the measurements in *Dictionnaire de physique* of 1789, but we have not been able to locate this source (it may refer to Lalande 1781). Ley 1964, p. 218, states that it was Montaigne who read the memoir to the Paris Academy in 1761, and that he "had repeatedly expressed skepticism toward the existence of a satellite of Venus." The first claim is wrong and the second lacks solid documentation.

[20]Lalande, who observed the 1761 transit from the Palais du Luxembourg, referred to the observations of Baudouin and Messier in Lalande 1763b (p. 84).

[21]Baudouin 1768a. In Baudouin 1768b he reported observations of two of Jupiter's moons, made in November 1761. The two communications were abstracted in Bernoulli 1771, pp. 156–158. Baudouin's articles and addresses read to the Paris Academy indicate that at the time he was considered with some respect by the French astronomical community.

> resemblance between the Moon and the satellite of Venus; for it is not only similar with respect to size, but also with respect to the distance from its primary planet.[22]

Baudouin realized that Montaigne's observations provided an opportunity to calculate the satellite's orbital elements, but his attempt to do so was not very promising. According to Jean-Étienne Montucla's history of mathematics, Lalande helped him with the mathematics necessary for the calculations.[23] Baudouin found the supposed moon to be moving in an orbit nearly perpendicular to the ecliptic, at a distance from the planet of 50–60 Venus radii and with a period of revolution of about 12 days.

As to the mass of Venus, since the time of Newton it had been known how to find the mass of a planet surrounded by one or more moons, if only approximately. In Book 3, Proposition 8, of the *Principia*, Newton had applied his theory of gravitation to find "the quantity of matter in the individual planets." He used the orbit of Venus to "weigh" the Sun and then, by means of Jupiter's outermost satellite (Callisto), to calculate the mass of Jupiter relative to the Sun. He also provided relative masses for Saturn (using "the Huygenian satellite," i.e. Titan) and the Earth. Compared with modern values, he got reasonably correct relative masses for the two large planets, while his value for the Earth was far off the mark, about twice its correct value.[24]

The method can be formulated by means of Kepler's third law, in Newton's formulation. According to this law, written in a modern version,

$$\frac{a^3}{T^2} = \frac{GM}{4\pi^2}\left(1 + \frac{m}{M}\right) \cong \frac{GM}{4\pi^2}$$

where M denotes the mass of the central body (e.g. the Sun) and m the mass of the planet, and a and T are the semimajor axis and the period of revolution of the planetary body, respectively. G is Newton's constant of gravitation. Applied to the two systems of Venus and its satellite, and the Earth and the Moon, we get

$$\frac{M_V}{M_E} \cong \left(\frac{a_s}{a_m}\right)^3 \left(\frac{T_m}{T_s}\right)^2 . \qquad (*)$$

Here, M_V = mass of Venus, M_E = mass of Earth, a_s = distance Venus-satellite, a_m = distance Earth-Moon, and T_m and T_s denote the periods of revolution of the

[22] Baudouin 1761c, p. 24.

[23] Montucla 1802, p. 16. The author must have known, for most of the volume was written by Lalande, who also served as an editor. (Montucla died in 1799, before his work was completed.)

[24] Newton 1999 (based on the third edition of 1726), pp. 812–813. See also I. Bernard Cohen's clear exposition on pp. 218–228 and the recalculation in Garisto 1991. Newton found that the mass of the Sun was 1,067 times that of Jupiter and 3,021 times that of Saturn; the modern values are 1,047 and 3,498, respectively. For the Earth, Newton got 169,282, while the present value is 332,946.

Moon and the Venus satellite, respectively. Baudouin adopted the ratios $a_s/a_m = 50/60$ and $T_m/T_s = 273/120$, but instead of inserting them in equation (*) – which would have led to $M_V/M_E \cong 3$ – he introduced in his calculation the ratio between the radii of Venus and the Earth. This ratio was usually taken to be slightly less than 1, but Baudouin adopted the value $R_V/R_E \cong (64.2/93) = 0.69$, which figure he got from a book of the French astronomer Pierre Charles Lemonnier, the *Institutions astronomiques* of 1746. Lemonnier gave two different values for the radius of the Sun relative to the radius of the Earth and Venus, respectively. According to one set of data, $R_S/R_E = 100$ and $R_S/R_V = 83.3$, and according to the other the values were 64.2 and 93. He believed the last set of data was the most precise.[25]

By multiplying $(a_s/a_m)^3(T_m/T_s)^2$ with the cube of R_V/R_E Baudouin obtained the value 0.98, which he mistakenly thought was the mass of Venus relative to the mass of the Earth: "I find in this way the figure 0.98, which is very close to 1; it then becomes clear that the mass of Venus is almost the same as that of the Earth, for the difference of 1/50 in calculations of this sort becomes negligible."[26] Had Baudouin used his data correctly, he would have got $M_V/M_E \cong 3$ and a density of Venus nine times the density of the Earth. Whatever the obscurity of his reasoning, he maintained that the mass of Venus was very close to that of the Earth.[27]

Montaigne's observations and the reports of Baudouin, delivered to the Académie des Sciences, attracted brief attention among the distinguished Paris astronomers. On behalf of the Royal Academy, Lalande added a supportive postscript to the first of Baudouin's memoirs in which he praised the work of Montaigne and Baudouin, describing it as a most important discovery. As a further sign of official approval, the postscript was certified by Grandjean de Fouchy, the Academy's perpetual secretary.[28] After Baudouin had reported on the last of Montaigne's observations, which he did on 26 May 1761, Lacaille and Lalande issued another certificate formulated as follows:

> We have examined, by order of the Academy, the remarks of M. Baudouin on a new observation of the satellite of Venus, made at Limoges the 11th of May by M. Montaigne. This fourth observation, of great importance for the theory of the satellite, has shown that its revolution must be longer than appeared by the first three observations.

[25] Lemonnier 1746, p. 558. Contrary to other observers, William Herschel found that Venus was a little larger than the Earth (Herschel 1912, vol. 1, p. 450).

[26] Baudouin 1761c, p. 48.

[27] Ibid., pp. 47–48. See also Lynn 1887a. It is hard to understand how Baudouin could have made this elementary mistake, especially if he was assisted by Lalande.

[28] Baudouin 1761a, pp. 23–25.

> M. Baudouin believes it may be fixed at 12 days; as to its distance, it appears to him to be 50 semidiameters of Venus; whence he infers that the mass of Venus is equal to that of the Earth. This mass of Venus is a very essential element to astronomy, as it enters into many computations, and produces different phenomena. But although M. Baudouin holds back in order to report many more observations about what is mentioned above, we consider this second memoir as an essential continuation of the first, and we believe it worthy of being printed.[29]

Although the two distinguished Paris astronomers refrained from accepting the discovery claim, with their statement the hypothesis of the satellite of Venus had received a kind of official authorization. Not only did Lalande express guarded support of the satellite in 1761, he continued to do so for several years, even after the case had been dismissed by the majority of astronomers. The positive attitude is evident from his detailed exposition in the *Encyclopédie* of 1778–82, which ends with an evaluation of the orbit of the satellite such as measured by Montaigne. According to Lalande, "This position perpendicular to the ecliptic, instead of being a ground for rejecting the existence of the satellite, seems to establish it with even more certainty, if one compares this phenomenon with what we know about the revolution of Venus around her axis."[30]

At about the same time Lalande contributed an article on Venus' moon to the eighth edition of the *Dictionnaire de physique* in which he presented the moon as a reality: "The year of 1761 will be celebrated in astronomy because of the discovery, made on 3 May, of a satellite of Venus." Referring to Baudouin's "very interesting memoir," he went through the calculations which led to the ratio of the mass of the Sun to that of Venus. Lalande's reasoning was, like Baudouin's, based on Kepler's third law, but instead of using the systems Venus-satellite and Earth-Moon he used the systems Venus-satellite and Sun-Venus. Then the version of equation (*) becomes

$$\frac{M_V}{M_\otimes} \cong \left(\frac{a_s}{a_V}\right)^3 \left(\frac{T_V}{T_s}\right)^2 \qquad (**)$$

where the symbol $\otimes$ denotes the mass of the Sun. For the satellite's period of revolution Lalande used the value $T_s = 223$ hours and for the ratio a_V/a_S his choice was 241. He found in this way $M_\otimes/M_V$, which he compared with $M_\otimes/M_E$, for which he used the too small value 207,197. His result, very different from the

[29] Baudouin 1761b, pp. 15–16, followed by a certification by Grandjean de Fouchy (1707–1788). Most of the passage appeared in English translation in Hutton 1795, vol. 2, p. 649. Neither of the two postscripts were included in the German translation. According to Hell 1792, p. 116, Lacaille was opposed to Baudouin's memoir and its claim of a Venus moon. He was, so Hell suggested, fooled or forced to support the publication of the work.

[30] Encyclopédie II, p. 260. Lalande's argument is not very clear.

one obtained by Baudouin, was this: "The mass of the Sun relative to the mass of Venus is 23,946 : 1; thus, the mass of Venus is about 8 times as great as that of the Earth, for we have demonstrated ... that the mass of the Sun relative to the mass of the Earth is 207,194 : 1."[31] Actually the ratio is 8.65, closer to 9 than to 8. Lalande did not comment on the discrepancy between his value of the mass of Venus and the one obtained by Baudouin. Nor did he try to estimate the mass density of Venus, in the manner Baudouin had done. Had he done so, and adopted Lemonnier's value of $R_V/R_E = 0.69$, he would have arrived at the incredible value of $\rho_V/\rho_E = 26.3$. (Had he used $R_V/R_E = 0.97$, as Lambert did in 1775, his result would have been $\rho_V/\rho_E = 9.5$.)

It is remarkable that Lalande considered Venus' moon in a favourable light for such a long time. By the early 1780s he may have been the only astronomer of distinction who did not dismiss the satellite. The reason for his positive attitude is not very clear, nor is it clear why he eventually lost faith in the existence of a satellite of Venus. In the second edition of his *Abrégé d'astronomie*, published in 1795, he mentioned explicitly that he no longer believed in the satellite: "The vain attempts that I, as well as several other astronomers, made to experience it has persuaded me that it is an optical illusion created in the glasses of the telescope; this is what Hell thought at the end of his Ephemerides of 1766 and Boscovich in his fifth dissertation of optics."[32] In the *Bibliographie astronomique*, completed in 1802, Lalande said about Baudouin that he "has often been of use to the astronomers." But he also wrote of the memoirs of 1761: "These memoirs were read to the Academy on the occasion of the observations that M. Montaigne thought to have made of a satellite of Venus; however, it is now recognized that it was an illusion."[33] It was common to dismiss the satellite of Venus with the argument that it had not been observed for several years. This was also Lalande's argument, but why did it take him thirty years or so to reach the conclusion?

Montaigne's true hunting ground was the world of comets, not the planets or the satellites. As a postscript to the observation claim of May 1761 it is relevant to point out that on 18 October 1780 he discovered a new comet, immediately communicating the news to Charles Messier in Paris. However, the Parisian comet hunter was unconvinced, as he failed to see the comet himself; he suggested that Montaigne might have mistaken a small star-cluster for it.[34] Only many years later, in 1801, did Heinrich Olbers reveal that he had seen the comet at the same time as

[31] Paulian 1781, pp. 336–337. The articles in the dictionary appeared anonymously, but it is most likely that the author of the entry on Venus' moon was Lalande.

[32] Lalande 1795, p. 322. On the arguments of Hell and Boscovich, see section 4.2.

[33] Lalande 1803, p. 477.

[34] Messier 1784, which includes excerpts of letters from Montaigne to Messier (1730–1817). For Montaigne's letters to Messier, dating from 1770, 1772 and 1780, see Bigourdan 1904.

Montaigne. Although the German astronomer came to the support of Montaigne, he did not fail to note that the reputation of the Frenchman was somewhat tainted by his earlier observation claim of the Venus moon. "It would be quite unjustified to completely dismiss a comet on the ground that it has only been seen by a single person," Bode wrote. "Certainly, Montaigne has also observed a Venus satellite that does not exist; but it is known that an optical phantom has deceived several able astronomers to see the alleged satellite. Montaigne was however experienced in comet observations, and he first discovered the comets of 1772 and 1774, both of which Messier found from his communications."[35]

Unknown to the astronomical community in the 1760s, a Jewish amateur astronomer in Crefeld, Germany, thought to have observed the Venus moon during the transit. Abraham Scheuten reported his observation to Lambert in letters of 1775, first anonymously and later by name. Contrary to most other observers, he was "completely convinced" that the moon existed and that he had proved it. He did not explain why he had remained silent for so long. According to Scheuten's observation notes from 6 June 1761: "This morning at $5\frac{1}{2}$ I saw Venus in the Sun. Because of the clouds that are here, it was not possible to make observations from 8 o'clock to 12 o'clock. At 12 o'clock I saw Venus and its small moon in the middle of the solar disk. At 3 o'clock it was near the limb."[36] Scheuten had no doubt that the object he had seen, and which he estimated to be about a quarter of the size of Venus, was the long-sought satellite. His telescope was not equipped with a micrometer, but he estimated that the moon passed the radius of the Sun in three hours, faster than Venus itself. Interestingly, the moon remained visible even after Venus had passed off the Sun.

To the same category may perhaps be counted an observation made in St. Neot in Huntingdonshire by an anonymous Englishman who reported to a London journal that he had observed the Venus transit and on this occasion noted an object that might perhaps be a moon:

> This morning as I was observing the transit, I perceived a phenomenon, which by its motion appeared to move in a curve different from any spots I had ever before discovered in the sun. An idea occurred to me, that it was the secondary planet to Venus: for it plainly appeared to attend its primary as the centre of its motion; and be help of my telescope I could perceive it to make near the same transit as the planet Venus but nearer the ecliptic. End of the transit of Venus 31 minutes past eight, and the end of the secondary 9 minutes past nine

[35] Olbers 1804, p. 176. See also Lynn 1887b.

[36] Letters of 14 November and 28 December 1775, reproduced in Lambert 1776, pp. 186–188. See also Stroobant 1887a, pp. 41–42.

in the morning, apparent time?[37]

A similar observation by a Danish amateur astronomer will be referred to in section 3.3.

3.2 Observations and non-observations

In the evaluation of the existence of a controversial object, non-observations may be of no less importance than observations. If there are reasons to expect the object to be seen, but it does not actually happen, then the non-observation counts as evidence against its existence. And if there are many non-observations of this kind, they may well overrule the positive evidence established by a few observations.

Among the wealth of communications that appeared in connection with the 1761 transit observations – more than 120 papers and reports were published – only a few included mention of a failure to observe Venus' satellite. It is obvious from the communications that some of the astronomers looked for the satellite, but it is impossible to say how many. In addition to the few who did mention the missing Venus moon, there were probably others who, not having seen it, just chose not to mention it. They may not have found it worth-while to report the non-observation of the hypothetical and controversial moon. For example, volume 52 of *Philosophical Transactions*, covering the years 1761–62, included 21 articles on the Venus transit. None of the articles reported that a satellite of Venus had actually been spotted, but three authors (an Englishman, a Swede and a Frenchman) mentioned that they had looked in vain for the satellite.

We know of six astronomers belonging to this category, but there may have been several more. In France, Nicolas-Louis de Lacaille, one of the chief architects behind the French transit project, was aware of the opportunity. However, "We did not see the appearance of the satellite on the Sun, neither on the 5th in the evening nor on the 6th until 3 p.m."[38] Also César-François Cassini de Thury, the son of Jacques Cassini and better known as Cassini III, referred to the missing moon. In his report on observations made at an observatory in Vienna belonging to the Jesuit mathematician and astronomer Joseph Liesganig, he wrote: "During all the observations I searched for the satellite that had been announced to appear in front of the Sun, but I could see nothing."[39] Another French astronomer, Alexandre-Guy

[37]Mentioned in Encyclopédie II, p. 259. The source is the *London Chronicle or Universal Evening Post* 9, no. 699 (16–18 June 1761), here quoted from Haase 1863–69, p. 11. See also Stroobant 1887a, p. 43. The observation in St. Neot was reported to the French-speaking world in an article of August the same year in the *Journal Étranger*.

[38]Lacaille 1763, p. 78.

[39]Cassini de Thury 1763, p. 412. J. L. Liesganig (1719–1799) served as astronomer and professor of mathematics at the Collegio Viennensi until 1773, when the Jesuit order was dissolved. Present

Pingré, was sent to the island of Rodrigues in the Indian Ocean to observe the transit. In his report to the Academy in Paris, he wrote: "I have seen no satellite of this planet; nor was Mr. [Denis] Thuillier, professor of mathematics, and appointed to assist me, by the King and the Academy, luckier than myself."[40]

Bengt (or Benedict) Ferner, the professor of astronomy in Uppsala, Sweden, was on an extended study tour abroad, and spent most of the year 1761 in Paris. He wrote about his observations in the Chateau de la Muette near Paris to Thomas Birch, the secretary of the Royal Society in London. In a postscript he commented: "I hope Mons. Baudouin's pieces upon the satellite of Venus is [*sic*] come to your hands. Notwithstanding all the care taken here, to discover this satellite upon the disk of the sun, on the 6th past, we could see nothing of it."[41]

In England, Samuel Dunn, a teacher and amateur scientist, observed the transit at Chelsea with a 6-foot reflector: "I carefully examined the sun's disk, to discover a satellite of Venus, but saw none." In a footnote he added: "After the transit, till two o'clock afternoon the same day, I continued observing the disk with this telescope, but saw no satellite pass over the sun."[42] Likewise, William Chapple and a friend observed the transit from Powderham Castle near Exeter, trying to catch a glimpse of the rumored satellite of Venus. But they only found two sunspots which did not follow the motion of the planet.[43] The Oxford professor of geometry Nathanael Bliss reported his observation of the Venus transit in a lengthy letter to the president of the Royal Society. Although he did not mention the satellite of Venus, on his drawing of the Sun's disk there appear three distinct dots, two of them close together and near the rim and the third just outside the disk.[44]

The news of Venus' alleged satellite had traveled as far as to America, where John Winthrop, professor of mathematics and natural philosophy at Harvard College, observed the transit. "I viewed the sun with great attention ..., in hopes to find a satellite of Venus; but in Vain. There were several spots then of the sun; but

---

at the observations in Vienna was also Maximilian Hell.

[40]Pingré 1763, p. 376. On Pingré's (1711–1796) adventurous voyage to Rodrigues and his transit observations, see Woolf 1959, pp. 98–115.

[41]Ferner 1763, p. 225, who made his observations in company with Jean Paul Grandjean de Fouchy, secretary of the Académie des Sciences from 1743 to 1776. The letter is dated 20 June 1761.

[42]Dunn 1763, p. 189. Samuel Dunn (1723–1794) wrote several tracts on astronomy, nautical science and instruments. In 1769 he was invited by Nevil Maskelyne (1732–1811), the Astronomer Royal, to observe the transit of Venus, and he was one of the first to observe evidence for an atmosphere on the planet (Meadows 1966).

[43]Chapple 1761a. In Chapple 1761b, he revised his data for the transit, without mentioning the satellite of Venus.

[44]Bliss 1762, figure on p. 244.

none that I saw could be a satellite."[45] Most likely, several amateur astronomers looked for the Venus moon in 1761, but without publishing their results. In a few cases we know about these observations because they were published at a later date. Scheuten thought that he had actually spotted the satellite, and so did a Danish observer, F. Artzt, whom we will meet in the next section. Johann Caspar Staudacher, an amateur astronomer from Nuremberg who specialized in sunspots, believed in the existence of the companion of Venus and on 6 June 1761 he looked eagerly for it with a 4-foot telescope. The result was disappointing: "No satellite was visible, but presumably it had already passed during the night or it stayed behind the body of the planet."[46]

The interest in Venus' moon greatly diminished during the next Venus transit of 1769, when we only know of one astronomer searching for the object (but, again, there may have been more). William Wales, a somewhat obscure British astronomer, was instructed by Nevil Maskelyne, the fifth Astronomer Royal, to make an expedition to northern Canada to observe the transit. Together with his assistant Joseph Dymond, Wales established an observatory at a place near the Churchill River in Hudson's Bay. When the day of the transit came, the two astronomers carefully observed the event. In their report to the Royal Society they noted that they had looked in vain for Venus' satellite: "We saw nothing like the appearance of an atmosphere round Venus ... either at the beginning, end, or during the time of the transit: nor could we see any thing of a satellite; though we looked for it several times."[47]

Although the many observations of 1761 had failed to confirm the existence of the Venus moon, a few years later claims of confirmation did appear, first in Copenhagen and slightly later in Auxerre, south of Paris, where a councillor by the name of Montbarron observed the satellite in late March 1764. He could not possibly have known of the Danish observations, made less than two weeks earlier and not made public until much later. Montbarron's observations were reported to Hell in a letter of 16 June 1764 from Messier, the famous comet hunter and compiler of celestial objects:

> While observing Venus with a 32-foot Gregorian telescope on 15 March 1764 at 7 h in the evening, Montbarron saw a small star at the dark side of Venus which formed an eastern angle of about 60°with

[45] Winthrop 1764, p. 283. Woolf 1959, pp. 93–94. On John Winthrop (1715–1779) as an astronomer, see Brasch 1916.

[46] Quoted in Wolf 1857, p. 276. Over a period of forty years, Staudacher (1731–1792) made observations of sunspots. These and other of his astronomical data were later published by another Nuremberg astronomer, Johann Woeckel (1807–1849). See Woeckel 1846.

[47] Wales and Dymond 1770. On the expedition to Canada, see Metz 2006. On Wales' astronomical and nautical voyages, see Orchiston and Howse 1998.

> the vertical. On the 28th of the same month, at $7\frac{1}{2}$ h in the evening, Montbarron saw the same small star near Venus, similar to the one observed earlier, forming a western angle of nearly 15°with the vertical. The following day, i.e. on 29 March, he saw the same star, although Venus was then covered by thin clouds. After this last observation he did not succeed to see the star again, and that in spite of having looked for it several times with the same telescope. All observations prove that the star has not been a fixed star, since these scintillate and are less localized. Yet, Montbarron asserts that he could not distinguish any phase in this star.[48]

Montbarron did not mention the distance between Venus and the "star." Both Messier and Hell looked for the Venus moon in the same period, in a series of observations from March to June 1764, but with no positive result. While Hell concluded that no satellite existed, and that earlier observation claims were based on optical illusions, Messier may have been less certain. At any rate, on the 8th of March 1766 he employed his excellent Gregorian telescope, which magnified objects about 110 times, "to look for the satellite of Venus, which for some years has been talked of." Although he failed to see the satellite, he did see "at some distance of Venus, a nebulosity of a small extent, with a luminous center."[49] The nebulosity, he realized, was a new comet. At some time after 1805, Messier wrote by hand a manuscript, *Notices des mes comètes*, in which he summarized for himself all the 44 comets he had observed. In this manuscript he wrote about the observation of March 1766: "Comet discovered on March 8, when searching the satellite which has been pretended to be seen at Venus; for this research I employed an excellent Gregorian telescope of 30 inches focal length, lent by M. le President de Saron, the large mirror of 6 inches diameter, and a very good achromatic refractor of 5 feet focal length; I did not see anything of the satellite."[50]

3.3 The Danish Connection

In 1761 the Venus moon was also observed in Copenhagen, at the "Round Tower" observatory established in 1642 as the first but one university observatory in the world (the first was in Leiden, dating from 1633). However, the event was not

[48]Reported in Hell 1765, p. 26, here quoted from the German translation in Haase 1863–69, pp. 252–253. A briefer account of Montbarron's observations was brought in Encyclopédie II, p. 259. Montbarron is not mentioned in Poggendorff 1863, Hockey 2007 or other standard sources.

[49]Messier 1767, p. 57.

[50]The handwritten manuscript is in the possession of the Observatoire de Paris. An English translation can be found online at
`http://seds.lpl.arizona.edu/messier/xtra/history/notes-c.html`.

publicly reported. The professor of astronomy and director of the observatory, Christian Horrebow, worked on stellar parallax determination and also made systematic observations of sunspots, including their variation in time. Although he did not discover the 11-year cycle, these data are today recognized to be of historical value.[51] He was engaged in the European Venus transit project, but the data from Copenhagen and Norway were of limited importance as they were based on imprecise timekeeping. Contrary to the Swedish astronomers, the Danes made no efforts to disseminate their observations by means of articles to the *Philosophical Transactions* or other widely read scientific journals.[52] Horrebow communicated the Copenhagen observations in the form of a Latin treatise and also a communication, in Danish, to the Royal Danish Academy of Sciences and Letters. Lalande, who had been informed by the French ambassador in Copenhagen, published a summary report on the Danish observations as well as those made by the young observer Thomas Bugge in Trondheim, Norway.[53]

Although Horrebow did not see Venus' moon in his observations of 1761, he was aware of its possible existence and interested in the phenomenon. Thus, in a dissertation on the Venus transit presented on 4 June, a kind of preparation for the observations to be made two days later, he referred to the observations of Cassini in the late seventeenth century, and also to "a certain report in England" (i.e., to Short's paper of 1740). He urged paying attention to the question of a possible satellite.[54] Horrebow also mentioned the satellite in another dissertation which he read at the University of Copenhagen on 28 July 1761 and in which he presented the recently found transit data: "We do not dare deny that Venus has a satellite. This real satellite, very different in nature from the other satellites in our known planetary system, is probably truly seen."[55]

Among Horrebow's trusted assistants at the Copenhagen Observatory was Peder Roedkiær, who happened to observe what he thought was the Venus moon after the transit had occurred, between 28 June and 1 December 1761. During these observations he observed several times the moon of Venus and recorded data

[51] Hoyt and Schatten 1995. The 11-year sunspot cycle was discovered about 1840 by Heinrich Schwabe (1789–1875), a pharmacist and amateur astronomer in Dessau, Germany, while looking for evidence of an intramercurial planet.

[52] Christian Horrebow was only appointed full professor in 1764, at the death of his father Peder Nielsen Horrebow (1679–1764) who held the chair 1713–64 (and had earlier worked as the assistant of Ole Rømer). However, since 1753 the younger Horrebow had served as director of the observatory and *de facto* as professor of astronomy. On Christian Horrebow, see Moesgaard 1972, who fails to mention his work on Venus and its satellite. On the Danish transit observations of 1761 and 1769, see also Pedersen 1992, pp. 102–104, and Nielsen 1957a (in Danish).

[53] Horrebow 1761a; Horrebow 1765c; Lalande 1763a.

[54] Horrebow 1761b, p. 20.

[55] Horrebow 1761a, unpaginated introduction.

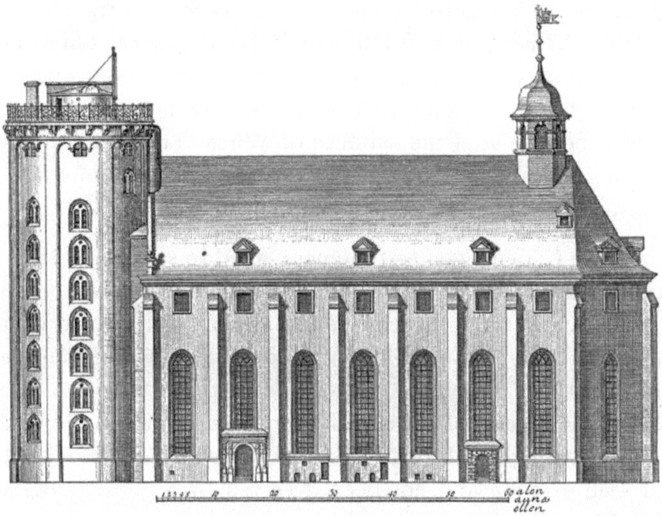

Figure 3.6: The Round Tower in Copenhagen with its observatory on the top, from where Christian Horrebow and his team of observers made their Venus observations 1761–68. Although the observation tower was built in conjunction with *Treenighedskirken* (the Trinity Church), it never functioned as a church tower. Illustration from the architect Laurids de Thura's *Hafnia hodierna* of 1748.

for its position relative to the planet.[56] As pointed out by Hans Schjellerup, who published the observations in 1882, it is a puzzle why Horrebow and Roedkiær chose to ignore them. The best answer is probably that Horrebow was uncertain if his assistant had really seen a satellite of Venus. The Copenhagen astronomers published observations made in 1764, but neither at this nor any other occasion did they refer to the data obtained three years earlier. Roedkiær initially thought that he might have seen the satellite of Venus only assisted by a quadrant. However, Horrebow and his other assistants were unable to confirm this first observation:

> While observing Venus with the quadrant Roedkiaer saw some whiteness which followed Venus. He found the distance between it and

[56]The observations of 1761 remained unknown until 1882, when the Copenhagen astronomer Hans C. F. C. Schellerup (1827–1887) reproduced them in their original Latin from the old observations ledger. See Schjellerup 1882, who however seems to have been unaware of Horrebow's two dissertations of 1761. The observations of Roedkiær and Horrebow are not well known in the history of science. Pedersen 1992, p. 80, briefly notes that Roedkiær published "a report of an observation of the (alleged) satellite of Venus," but that is all.

> the upper rim of Venus to be 0.66, and he observed a transit of 11"between it and Venus. After that he saw it again by means of a telescope of 17′, and because its appearance was sickel-shaped, not as pronounced as that of Venus but shining with almost half its face, the observer surmised that he had seen the satellite of Venus. The others of us could not see this whiteness even though we observed Venus often, with the quadrant, the meridian circle and the 17 feet telescope.[57]

Although the Copenhagen astronomers failed to see the satellite of Venus in connection with the transit, unbeknownst to them another Dane spotted the object on 6 June. Friedrich Artzt, a secretary and amateur astronomer, observed Venus passing in front of the Sun with a 3-foot reflecting telescope from a place on Zealand called Gundersløvholm. When the planet had reached the centre of the Sun, he saw a smaller globe entering the disk and apparently following Venus throughout the transit. Recollecting his observation many years later, Artzt stated confidently that he had seen a moon and not been deceived by a ghost image produced in the telescope. Because, "When Venus had passed the Sun, the deception ought to have disappeared with it, but this did not happen; the small globe or moon was at the time near the middle of the solar disk and continued its course ... along the same line, and finally it left the Sun at the same place as Venus."[58] The satellite was behind the planet with about four hours, and Artzt estimated that its radius was perhaps one-fifth of that of Venus.

Artzt was aware that transit observations of the Venus moon might sometimes be due to sunspots, but not in his case. He saw five spots on the Sun, including one moving in the same orbit as the moon, but this spot "was for a brief while hidden by it [the moon], which is another proof of the presence of the satellite." He did not explain why he waited more than fifty years to report his observation, which seems to have remained unknown until 1813, also among the astronomers in Copenhagen.

To return to Copenhagen, on the evenings of 3 and 4 March 1764, Roedkiær saw a faint luminous object near Venus of size about one-fourth of that of the planet, i.e., the same size as Montaigne had reported. Its distance from the planet decreased from about three-fourth to one half of the Venus diameter. The observation ledger of 4 March included a rough drawing of the configurations of Venus and its satellite, and reported:

> This evening ... Roedkiær again saw the Satellite of Venus. Its distance to the left from Venus was $\frac{1}{2}$ of Venus' diameter. Its centre

[57] Observation report of 28 june 1761, in Schellerup 1882, p. 165. This and the subsequent translations from Latin have kindly been made by Henk Bos.

[58] Artzt 1813, pp. 453–454.

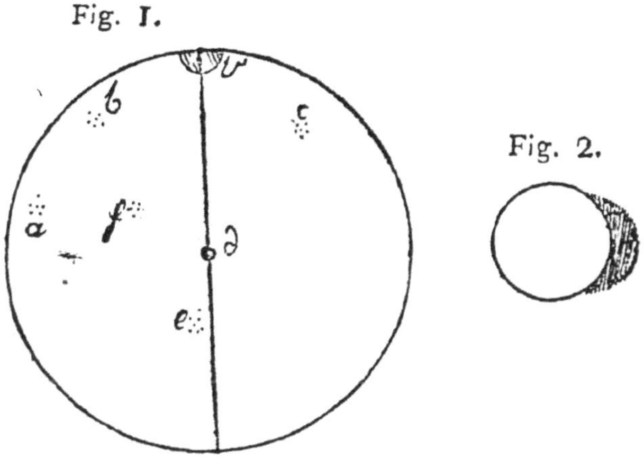

Figure 3.7: Friedrich Artzt's drawing of 1813 of the satellite of Venus, as he saw it during the 1761 transit. The large circle to the left is the Sun's disc with Venus (v) at the top. The satellite, close to the centre, is denoted with a ∂. The smaller figure to the right is Artzt's impression of Venus and the "fluid and transparent substance" which he saw at the southern pole of the planet. The observation claim went unnoticed, both nationally and internationally.

> made with Venus' centre an angle of about half a right angle: it appeared higher than Venus' centre in the telescope. He also could very well distinguish its phase which conformed to Venus' phase. ... That it was a satellite was clear primarily because both the diameters of Venus and the satellite were enlarged noticeably (by the telescope of 14 feet as compared with the telescope of 9 $\frac{1}{2}$ feet), which applied to none of the fixed stars.[59]

Roedkiær was convinced that what he had seen was Venus' moon and not merely an optical reflection. As he made clear in a communication to the *Skrifter* of the Royal Danish Academy of Sciences and Letters, he believed he had confirmed Montaigne's earlier observation such as discussed by Baudouin. "I can now assure that there no longer is any doubt about the satellite of Venus," he wrote.[60] Christian Horrebow shared his assistant's belief and communicated the findings

[59] Schjellerup 1882, pp. 166–167.
[60] Roedkiær 1765. See also Horrebow 1765c.

to the Royal Danish Academy on 9 March 1764. Horrebow had himself looked for the satellite, but in vain, and also Johan Ahl, a Swedish instrument maker who had emigrated to Copenhagen, had failed to see the satellite of Venus.

On this same night, the satellite was seen again by Roedkiær, this time in company with Christian Boserup and Peder Horrebow (a brother of Christian Horrebow). On 11 March Christian Horrebow finally succeeded in seeing the moon. He was much impressed:

> I have never before seen a spectacle in the heavens which has captivated me more; I thought that I truly saw the satellite of Venus and felt happy in my heart that I now saw that the Lord had provided the inhabitants of Venus with a satellite, just as ours. I sought to establish in many ways whether this weakly luminating body might be a deceptive reflection in the telescope, but ... [reached the conclusion] that the light must really be the Venus satellite. ... To describe this observation more closely I know of no better way than to refer to precisely the expressions that Mr. Cassini uses when he describes his observations of 25 January 1672 and 28 August 1686. All of these fit closely with the ones here observed, and thus our observation might be considered a perfect repetition of the ones reported by Cassini.[61]

In spite of his obvious fascination and joy of having made an important discovery, Horrebow was by no means naive or uncritical. An experienced astronomer, he knew well the dangers of mistaking spurious light signals for real objects. He argued that reflections were well known to astronomers and that they were used to take care of them. Moreover, why would the moon-imitating reflections only turn up in the few cases when a possible satellite had been seen, and not in all the other observations of Venus?

> To be more certain, on the same evening when I saw Venus' satellite I turned the telescope towards Jupiter and Saturn, and I saw them both very distinctly and precisely ... without any indication at all of a false light in the telescope. What is more, during the observations I turned the telescope in a variety of ways, and yet the position of the satellite relative to Venus always remained fixed. In addition, a couple of times I let Venus pass through the tube, from beginning to end, and the satellite followed its primary planet all the time, just as it should; had it been a reflection, it would sometimes have disappeared. In the case where I arranged the telescope in such a way that Venus was just outside it, I could still see the weak light of the lone satellite.[62]

[61] Horrebow 1765c, p. 402.
[62] Ibid., p. 403.

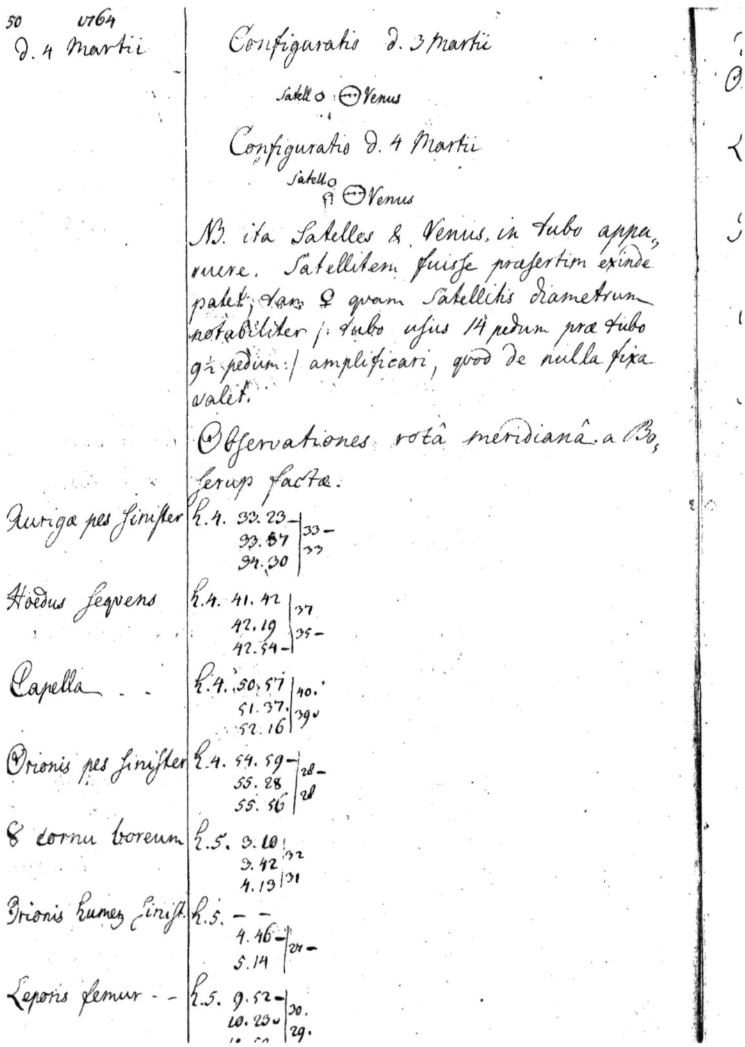
50 1764
d. 4 Martii
Configuratio d. 3 Martii
Satell. Venus
Configuratio d. 4 Martii
Satelles Venus
NB. ita Satelles & Venus, in tubo apparuere. Satellitem fuisse praesertim exinde patet, dum ♀ quam Satellitis diametrum notabiliter (: tubo usus 14 pedum prae tubo 9½ pedum :) amplificari, quod de nulla fixa valet.
Observationes rotâ meridianâ a Boserup factae.
Aurigae pes sinister h.4. 33.23 33.57 34.30
Hoedus sequens h.4. 41.42 42.19 42.54
Capella h.4. 50.57 51.37 52.16
Orionis pes sinister h.4. 54.59 55.28 55.56
♉ cornu boreum h.5. 3.10 3.42 4.13
Orionis humerus sinist. h.5. 4.46 5.14
Leporis femur h.5. 9.52 10.23

Figure 3.8: From the Copenhagen Observatory ledger. On the top is Roedkiær's drawings of the Venus satellite as seen on 3 and 4 March 1764. Below the drawings of the configurations, he notes: "N.B. This is how the satellite and Venus appeared in the telescope." Courtesy: Institute of Science Studies, University of Aarhus.

Horrebow was also aware of the objection that what he had taken to be a satellite might be a fixed star, and he consequently argued that this could not possibly be the case. Not only did the supposed satellite look very different than a star, he had also observed it moving around Venus. Furthermore, none of the known fixed stars were in the area where he had seen the moon. He concluded his account by encouraging other astronomers to follow up the observations in Copenhagen and thus convincing themselves that the satellite of Venus really existed. The observations, he optimistically hoped, would free astronomers from the "fear and modesty" that had prevented them from presenting corroborative data.

In regard to Horrebow's confidence that he had made a major discovery, it is remarkable that he did not communicate the discovery claim to astronomers abroad or otherwise followed it up. As far as we know, it remained largely unnoticed in the international astronomical literature. The observations of Roedkiær and Horrebow were not mentioned by Lalande in his article in the *Encyclopédie*, nor did they appear in the *Philosophical Transactions* or the *Histoire et Mémoires* of the Paris Academy. Nonetheless, they were not completely unknown, as they were reported in some detail in the *Gazette Litéraire de l'Europe* of 18 April 1764. It was from this source that they found their way to Maximilian Hell's comprehensive *De satellite Veneris* of 1766 (see below).[63]

Nearly four years later, on the evening of 4 January 1768, Horrebow saw the moon one last time, now in company with his assistants Ole Nicolai Bützov and Ejolvor Johnsen (Roedkiær had died the previous year). Using a Dollond telescope, the three astronomers observed below Venus "a small light, certainly not a star (for there were stars in the telescope, which had a fully different appearance), and it stood at a distance from Venus of about one Venus diameter. Soon afterwards Venus was observed in the Islaean telescope of 12 feet." Horrebow believed that the new observation confirmed the hypothesis of a Venus satellite. "After an hour or three quarters of an hour that light which adhered to Venus appeared more to the right in the Dollian and more to the left in the astronomical telescope. Three observers observed this same phenomenon at Venus, C. H., O. B., and J.; all saw with certainty that this light was not a star, and were certain that the light was not an optical illusion, and they therefore surmised that perhaps it was a satellite of Venus."[64]

---

[63] A German translation of Hell's Latin version of the French account in the *Gazette Litéraire* (based on the original Danish) appears in Haase 1863–69, pp. 251–252.

[64] Schjellerup 1882, pp. 167–168. The "Islaean telescope" refers to a kind of refractor named after the French physicist and astronomer Joseph Nicolas Delisle (1688–1768). C. H. = Christian Horrebow; O. B. = Ole Bützov; J. = Ejolvor Johnsen. In company with Peder Horrebow junior (1728–1812), Bützov (1742–1784) went the following year to the northern parts of Norway to observe the Venus transit. During 1773–78 he was in charge of the small observatory at Vardø.

Although Horrebow's belief in Venus' moon seems to have remained unshaken, he did not make the 1768 observation public, not even in a paper to the proceedings of the local academy of science. He read papers before the Royal Danish Academy of Sciences on 18 February 1768 and 6 April 1770, in both cases on the transit of Venus. However, these were oral presentations only and it is unknown if he referred to the satellite observation of January 1768. Might he have been victim to the "fear and modesty" he had previously warned against? Or did he suspect that the observation might have been false after all? According to Maximilian Hell, who met Horrebow in Copenhagen in May 1768, the Danish astronomer did not believe in a satellite of Venus. Nor had he ever believed that his and Roedkiær's observations of 1764 justified the conclusion that Venus had an attendant! No, they were merely optical illusions, and Horrebow's "observations had been reported in the newspapers against his will." According to Hell:

> He had made a public statement in the Academy that what he saw was just an indefinite light which was neither round nor showed phases like Venus; it looked like a little cloud. He had also been convinced that it was an optical illusion when he had been unable to see the little cloud with a 18-foot telescope. On the other hand, he saw it with a less powerful telescope of $9\frac{1}{2}$ feet. He agreed with my opinion that also the observations of other astronomers had been optical illusions, such as I had explained in my treatise.[65]

Needless to say, Hell's report of the conversation in Copenhagen does not agree with either Horrebow's published account of 1764 or his unpublished comments on the observation of January 1768. It is hard to avoid the conclusion that Hell either misunderstood what Horrebow told him or deliberately misrepresented the opinion of the Danish astronomer. At any rate, as far as observations are concerned, Venus' satellite disappeared quietly and with no fanfare, both from Danish and international astronomy.

After the death of Christian Horrebow, the Copenhagen chair in astronomy was taken over by Thomas Bugge, an internationally oriented astronomer and geodesist. In a report to the Royal Danish Academy of 1781 on Herschel's discovery of Uranus he referred briefly to the "much-discussed satellite of Venus." He found it conspicuous that it had only been seen so rarely and irregularly and concluded that the observations were probably optical illusions, such as suggested by Hell. He was inclined to believe, he wrote, "that it was due to an optical illusion in the telescope, and the esteemed Vienna astronomer, Mr. Hell, has shown that in any telescope and at any planet, when the eye is in a certain position, there appears

[65]Hell 1792, p. 118. Hell was on his way to Norway to observe the Venus transit of 1769 (see section 4.2).

close to the planet a dioptrical ghost or a small imitation of the main planet." He further pointed out that Lambert had calculated the satellite to be visible in front of the Sun on 1 June 1777. "The Sun was carefully observed the entire day in Vienna, Berlin and Copenhagen, but not the slightest trace of a satellite was found; from this one can safely conclude that what has been seen and calculated to be Venus' satellite has not been a real celestial body."[66] As far as Bugge was concerned, the satellite just did not exist. At the time, his attitude was shared by the large majority of astronomers.

3.4 The Venus moon and enlightenment culture

What for a short period might have looked like a vindication of the old idea of a Venus moon turned out to be a false alarm. None of the observation claims of the 1760s succeeded in changing the majority view that most likely the satellite of Venus was a phantom. After all, it was hard to believe that the moon, if it were real, could have escaped unambiguous detection by the dozens of astronomers eagerly observing the transit of 1761. Nor was there any indication of it in connection with the next Venus transit eight years later, on 3 June 1769, which engaged an even larger number of astronomers. Whatever excitement there had been over the satellite, it quickly died out in lack of further support. Noting that the satellite had not been observed for a decade, Johann Lambert tersely remarked: "Perhaps the desire to look for it has diminished. When a satellite has not been observed for several consecutive years one easily tires of looking further for it."[67]

If the astronomers were reluctant to admit the existence of a Venus moon, the hypothesis was unreservedly supported by the Swiss enlightenment naturalist and philosophical writer Charles Bonnet. A pioneer of entomology and other parts of the biological sciences, Bonnet was strongly opposed to ideas of naturally generative processes. A main message of his *Considérations sur les corps organisés* from 1762 was that evolution only takes place as the unfolding of pre-existent forms, a doctrine known as preformationism. Bonnet made his public breakthrough with the celebrated *Contemplation de la nature* of 1764, a book that ran through several editions and was translated into German, Italian, English, Dutch, Danish and some other languages. In this popular and eloquently written work on natural history, he expounded the view that all beings in nature – including inorganic bodies – form a gradual scale from the lowest to the highest. According to the "principle of plenitude" there is no break in the continuity of existence – all entities that can exist, do exist.[68] The *Contemplation* dealt primarily with organic nature,

[66]Bugge 1783, p. 219. On the works of Hell and Lambert, see sections 4.2 and 4.3.

[67]Lambert 1775, p. 181.

[68]On Bonnet (1720–1793) and the principle of plenitude, see the classic account in Lovejoy

Figure 3.9: Charles Bonnet's *Contemplation de la nature*, first published in 1764, became a best-seller. It included a section in which the author advocated the existence of a satellite of Venus. At least implicitly, Bonnet's reasons were grounded in the physico-theological belief in a great chain of being.

but in his introductory chapter Bonnet also mentioned the astronomical bodies, including what were often called the secondary planets:

> Venus and the Earth have each their satellite. Undoubtedly, one day we will discover one at Mars. Jupiter has four [moons], Saturn has five in addition to its ring or luminous atmosphere, which may replace many small moons. Since it is nearly three hundred million miles away from the Sun, it would receive such a faint light if it were not for its

1964, first published in 1936.

> satellites and ring which reflect and amplify the light. We know seventeen planets [including moons] which help to make up our solar system; but we cannot rest assure that there are not more present. Since the time of the invention of the telescope their number has increased considerably; and perhaps it will increase even more when we get better tools and become more diligent and fortunate observers. This Venus satellite, glimpsed in the last century and again seen recently, heralds new conquests in astronomy. Not only was it reserved for modern astronomy to enrich our sky with new planets, it was also given to it to extend the boundaries of our solar system.[69]

Given the popularity of Bonnet's book, and recalling the extensive article in the *Encyclopédie*, acquaintance with the supposed Venus moon must have been widespread in cultured circles in the last decades of the eighteenth century.

The German translation of Bonnet's book was undertaken by Johann Daniel Titius, professor of physics in Wittenberg, who in his edition of 1766 inserted in the main text a section in which he presented for the first time what later became known as the Titius-Bode law or just Bode's law. This law, which exists in several versions, states that the relative mean distances of the planets from the Sun follow a mathematical series (for example, that $r = 0.4 + 0.3 \times 2^n$, where $n = -\infty$ for Mercury and $0, 1, 2, \ldots$ for the other planets). Titius suggested that there might exist a new planet in the space between Mars and Jupiter, although he described it as a secondary planet, namely a satellite of Mars. Reflecting on the large distance between Mars and Jupiter, "where at present neither a primary nor a secondary planet is to be seen," he wrote: "But should the Lord Architect have left that space empty? Never! Let us confidently wager that, without doubt, this place belongs to the as yet still undiscovered satellites of Mars; let us also add that perhaps Jupiter still has around itself some smaller ones that until now have not been sighted by any telescope."[70]

In a later translation of 1772, Titius added a footnote to Bonnet's reference to Venus' moon, noting that Bonnet's optimistic hope of 1764 had not yet been fulfilled: "The most recent observations of the passage of Venus across the disk of the Sun, made in all parts of the world, have failed to discover it [the satellite of

[69] Bonnet 1764, pp. 7–8. In a later edition, Bonnet added in a lengthy footnote a sketch of the history of the satellite of Venus from Cassini to Lambert, leaving the erroneous impression that Cassini, as well as Short and Lambert, firmly believed in the existence of the satellite (see the excerpt in Thornton 1804, p. 328).

[70] Bonnet 1766, pp. 7–8. Reproduced in Nieto 1972, plate IV. Readers of the German translation would believe that the law of planetary distances was due to Bonnet, as Titius (1729–1796) did not disclose the authorship of the inserted section. For details on the origin and historical development of the Titius-Bode law, see Jaki 1972a, Jaki 1972b and Nieto 1972.

Venus]. Although the case for it is likely, nonetheless it has not been confirmed by actual observations."[71]

In spite of the doubts that the lack of observational evidence cast upon Venus' alleged moon, in the 1760s it seems to have been the subject of considerable interest. An enthusiastic patron of science and culture, Frederick the Great of Prussia took it seriously enough that he proposed to name it after his admired friend Jean le Rond d'Alembert, who since 1746 had been a member of the Berlin Academy. The king wanted d'Alembert to become president of the Academy and never gave up his attempt to lure the eminent mathematician to Berlin.[72] Frederick the Great's flattering proposal to name the Venus moon after him may have been part of his maneuver to get him to the Prussian capital. However, d'Alembert declined his generous offers and remained in Paris. Perhaps he realized that it would not be much of an honour to have a non-existing moon named after him. Whatever his reasons, in a letter of 3 November 1764 he wrote: "It is not the first time there is a question of a Venus satellite, which H.M. [His Majesty] gives me the honour to talk to me about; and surely the Academy does not ignore it." He then gave a brief account of the history of the satellite, mentioning Fontana, Cassini and Short (but not Baudouin), and continued:

> It gives me too much of an honour to intend to baptize this new planet in my name; I am neither so great to be the satellite of Venus in the sky nor in so good health to be so on Earth; and I find myself so well in my little place which I occupy in this world to strive for one in the firmament. If one day a Mars satellite is found I know well what name I would give it, that of one of His Majesty's generals. With regard to Mercury, if it should be given the honour of a satellite, rather a maltôtier [financer or profiteer] or a courtier will furnish us with more names; but this god has already too many satellites on Earth to be anxious to have more.[73]

The enigmatic satellite was also known to another of the king's favourites, François-Marie Arouet, better known by his pen name Voltaire. Frederick courted the French author, wanting him to set down in Berlin as the brightest ornament of his new Academy. In a letter of 5 December 1742, the king flattered Voltaire

[71] Same title and publisher as Bonnet 1766, p. 7. Reproduced in Nieto 1972, plate V.

[72] For valuable background to Frederick II (1712–1786) as a philosopher-king and his re-establishment of the Berlin Academy in the 1740s, see Terrall 1990. See also Taton 1984, reprinted in Taton 2000, pp. 261–272.

[73] Frederic II, 1805, p. 19 and also in d'Alembert 1822, pp. 256–257. Part of the letter has been quoted in the earlier literature, e.g. in Schorr 1875 (p. 73), Wolf 1877 (p. 679) and Littrow 1886 (p. 101), but always without date and reference. Ley 1964, p. 219, states that King Frederick read Lambert's essay, a claim for which there is no documentation.

by calling him the satellite of Venus: "Si Paris est l'île de Cythère, vous êtes assurément le satellite de Vénus."[74] Although Voltaire did go to Berlin for a couple of years, his stay there was troubled and not very productive.

In a letter of 9 July 1761 d'Alembert referred to the satellite of Venus and the failure of observing it during the recent transit. "I do not know what has happened with the lackey of Venus," he told Voltaire. "I am afraid it cannot be a hired lackey which has ceased to stay with her for a long time, but rather that the said lackey has declined to follow his mistress during her passage over the Sun."[75] He further mentioned the great efforts of the French astronomers to observe the Venus transit from different places of the world, as far away as India and Siberia. As a result of these efforts and other observations d'Alembert had abandoned any hope that Venus might be provided with a lackey moon of her own.

Voltaire referred to the satellite in a polemical work published 1768, *Singularités de la nature*, which was principally concerned with philosophical questions related to embryology and natural history. It included attacks on the English naturalist and advocate of spontaneous generation John Turberville Needham, and also criticized Buffon, Maupertuis and other biological thinkers. In between these attacks, Voltaire digressed on astronomical matters. "It is nearly a century ago that it was believed that a satellite of Venus had been discovered," he wrote. "Since then, a famous English observer saw – or believed to have seen – this satellite, which is also believed to have been seen in France. However, the astronomers have seen nothing of it. It may exist – but we must wait and see."[76] Voltaire further suggested from reasons of analogy the existence of a satellite of Mars, but admitted that the suggestion lacked observational support. As mentioned in section 2.2, he had ventured the idea of a satellite of Mars earlier, in his *Micromégas* of 1752.

Although not an astronomer, the German philosopher Johann Gottfried Herder, who had studied with Immanuel Kant at Königsberg, had a lifelong interest in matters of astronomy. In an unpublished manuscript from 1765, *Anfangsgründe der Sternkunde*, he said, referring to Cassini and Short, that so far Venus was known to have a single satellite.[77] However, in a later theological work,

[74]From online edition of *Oeuvres Complète de Voltaire* (`http://www.voltaire-integral.com/`). In English: "If Paris is the island of Cythera, you are surely the satellite of Venus."

[75]d'Alembert 1822, p. 79.

[76]*Oeuvres Complète de Voltaire*, Mélanges VI. The title of Voltaire's work was *Des singularités de la nature par un académicien de Londres, de Bologne, de Pétersbourg, de Berlin etc.* See Roe 1985. On Voltaire and the sciences, see Perkins 1965, which mostly deals with his critical attitude to ideas of natural history.

[77]Quoted in Nisbet 1970, p. 141, which offers a detailed exposition of Herder's view on science and suggests that his ideas of astronomy were inspired by Kant. *Anfangsgründe der Sternkunde* was based on notes made by Liborius Bergmann, a student in Herder's class in the Cathedral school in Riga in 1765. Given that G. C. Reccard, the translator of Baudouin's memoirs on the

framed as a dialogue, he changed his mind. Philolaus, one of Herder's discussants in the dialogue, objected to the tendency of ascribing to God particular purposes with his creation. "From the register of divine intentions so much has been said and believed about the ring of Saturn, the Moon of the Earth and the satellite of Venus, that one had to retract it disgracefully when it was found that Venus has no satellite. . . ."[78] The shift in attitude may have been due to the lack of confirming evidence, but it may also reflect Herder's increasing dissatisfaction with natural theology and design arguments. By the 1780s he had moved closer to a broad natural religion. A correspondent of Lambert, Herder was well acquainted with his theory of the universe as described in the *Cosmologische Briefe*, which he called "a glory of the human intelligence."[79] In general his philosophy of history was inspired by the cosmological works of Kant and Lambert.

Given that Herder had studied under Kant, and the latter had an interest in Venus' moon (as indicated by his reading of Baudouin's memoir), it is quite possible that Herder became acquainted with the enigmatic satellite through Kant. In his later, so influential, cosmological treatise *Allgemeine Naturgeschichte und Theorie des Himmels*, anonymously published in 1755, Kant dealt in detail with Saturn and the origin of its ring. In fewer details he considered the circumstances under which a planet may obtain a satellite, concluding that only planets with large mass and at great distance from the Sun could be endowed with one or more satellites. "Jupiter and Saturn, the two greatest and also most distant planets, have most moons. To the Earth, which is much smaller than those, only one was allotted; and Mars, to which because of its distance some share in this advantage pertained, goes away empty because its mass is so small."[80] Kant did not mention Venus, but it followed from his reasoning that it would have no companion either.

Although the Venus moon did not appear in Fontenelle's celebrated *Entretiens sur la pluralité des mondes*, it did turn up in some of the later works of the same genre, that is, dialogues between a knowledgeable gentleman and a young inquiring woman. A London instrument maker and lecturer, Benjamin Martin was a writer of educational and philosophical works. In 1759 he published *The Young Gentleman and Lady's Philosophy*, a popular work framed as a dialogue between Cleonicus and his younger sister Euphrosyne. In the third edition of 1781, the young woman asks about Venus' satellite, "which has been of late so much the subject of Conversation, and which some of the Gentlemen present apprehended

Venus moon (cp. section 3.1), served as professor in Königsberg at the same time as Kant and Herder, one can imagine that the topic was discussed and widely known.

[78] Herder 1787, p. 35. See also Crowe 1999, p. 151. One of the minor planets (1989 UH7) is named after Herder.

[79] Palti 1999, p. 326, who argues the connection between Herder's philosophy and the cosmological views of Kant and Lambert.

[80] Kant 1981, pp. 131–132.

they saw on the solar Disk." Cleonicus offers as his opinion "that what the Gentlemen took for a Satellite in the Sun was only a solar Spot, and it is certain that what was published in the common News-papers was the same Kind of Mistake; for the Satellite having a similar Appearance with that of the Planet, ... must necessarily have been very easily seen on the Sun's Disk, should any such Thing been there." He elaborates:

> From the Accounts of Mr. Cassini, Mr. Short, and now of these French Philosophers, it is very certain there is something that must have appeared about the Planet in their Telescopes; but whether it be a Satellite of the same Nature and Kind with those which belong to the Earth, Jupiter, and Saturn, is a very great Question, because of the unusual Position of its Orbit, and of its very seldom and precarious Appearance. Were it only an opake Body, like another common Satellite, it would, like them, appear at all Times in a uniform Manner, more or less enlightened, but generally in the Form of a Crescent, like the primary Planet itself; and as Venus is so near to the Earth, at her inferior Conjunctions especially, it would be impossible not to have a frequent View of a common Satellite that attended her.[81]

Cleonicus admits that if the existence of the satellite is ever confirmed, "it will conduce not a little to the Advancement of Astronomy and natural Philosophy," for other reasons because it will yield an authoritative value of the mass of Venus. Before the couple proceed to other topics, Euphrosyne expresses her hope that "the Vigilance of Astronomers will one Day or other put that Matter out of all Doubt, and satisfy Posterity whether it be a real Satellite, or a mere *Ignis Fatuus* that now amuses them and perplexes their Curiosity."[82] Astronomers may not have cared, but in the 1780s the case of the satellite of Venus was far from a dead issue in popular and cultural circles.

Another work in the same literary genre, only more strongly permeated by physico-theology, was produced 1776-79 by the Dutch minister and science popularizer Johannes Florentius Martinet.[83] His often reprinted four-volume *Katechismus der Natuur* appeared in 1779 in a German translation. Martinet, a major exponent of natural theology and the great chain of being, may have been inspired

[81] Martin 1781, p. 263. *The Young Gentleman and Lady's Philosophy* first appeared in the *General Magazine of Arts and Sciences*, a journal edited and largely written by Martin (1704–1782) in the years 1755–65.

[82] Ibid., p. 264.

[83] Martinet 1779, where the Venus moon is discussed on pp. 212–214 and on p. 1 in the appendix. On Martinet (1729–1795) as an exponent of enlightenment physico-theology, see van der Wall 2004. In 1769 he was elected a member of the Dutch Society of Sciences (Hollandsche Maatschappij van Wetenschappen), founded in Haarlem in 1752.

Figure 3.10: Frontispiece of Benjamin Martin's *The Young Gentleman and Lady's Philosophy* (1759), showing a didactic conversation between brother (Cleonicus) and sister (Euphrosyne). Part of their conversation dealt with the satellite of Venus. Books like Martin's helped disseminate the hypothetical satellite to a broader public.

by Bonnet, whose *Contemplation de la nature* had recently been translated into Dutch. The fourth volume of the catechism of nature included a dialogue concerning the satellite of Venus, not unlike the one in Martin's book. However, Martinet's exposition was more historically informed and paid much attention to Scheuten's observation of 1761. He also referred favourably to Bonnet's opinion, which he seems to have shared. The popularity of the idea of a Venus moon is further illustrated by a Dutch work of 1791 which included a discussion of the subject in relation to Martinet's book. *Wis- Natuur- en Sterrekundige Briefwisseling* received a careful review in which the case of Venus' satellite was discussed in considerable detail.[84]

During the end of the eighteenth century, the astronomy-for-ladies genre had developed into a minor industry, with the satellite of Venus appearing in several of the works. As one more example, this time from Germany, consider Johann Heinrich Helmuth's *Anleitung zur Kenntnis des grossen Weltbaues*, characteristically subtitled *für Frauenzimmer in freundschaftlichen Briefen.* Helmuth, a theologian and author of works in natural and moral philosophy, explained the secrets of the skies through a correspondence between two young women and their tutor. The tutor (that is, Helmuth) told how he had failed to see the moon of Venus on 1 June 1777, the day that J. H. Lambert had predicted it should appear in front of the Sun. Referring to Hell's explanation in terms of optical illusions, he concluded that "it is pretty certain that Venus has no moon."[85] His attitude to the satellite of Mars was entirely different: "We find it very probable that Mars is surrounded by one or two moons," he wrote. The fact that no Martian moon had been observed, while the one of Venus had been seen several times, did not disturb the German theologian. His argument was this: "The splendid order of nature with respect to the distribution of moons should not be broken, and therefore we must continue to believe that the Creator has found it necessary to provide also Mars with one or more moons."[86] What Helmuth saw as "the splendid order of nature" was evidently that the exterior planets were all surrounded by moons, whereas the interior planets were not.

As we shall see, the moon of Venus continued to attract cultural and popular attention through the last decade of the eighteenth century and well into the following century. Before looking at the further fate of the satellite, from about 1780 to the late nineteenth century, we call attention to the attempts of some enlightenment astronomers to make sense of the conflicting observations from Fontana to Horrebow. It was generally agreed to dismiss these observations, but they could not be written off as dreams. The astronomers had seen something that looked like an attendant of the Cytherean planet. Had they seen a satellite of Venus? If not, what had they seen?

[84]Review in *Algemeene Vaderlandsche Letter-Oefeningen* 1792, pp. 97–101.

[85]Helmuth 1794 (second edition), p. 196. The first edition was published in 1791. Helmuth (1732–1813) was superintendent and minister in Calvörde in the dukedom of Braunschweig-Lüneburg. On Lambert's prediction, see section 4.3.

[86]Ibid., p. 287.

Chapter 4

Contemporary analysis and criticism

From the few and erratic observations of Venus' satellite it was impossible to conclude with any certainty whether the object existed or not. Theory offered no help, for neither physical nor astronomical theory had anything to say about the number and distribution of satellites in the solar system. If it was assumed that the companion of Venus really existed – and it just might – why was it seen only so rarely and irregularly? Conversely, if it was assumed that it did not exist, such as the majority of astronomers thought, how could the observation claims be accounted for? After all, something had definitely been seen, real or not. During the 1760s and 1770s, when the non-existence of the moon became the favoured view, basically three explanatory accounts were developed to address the questions. One was due to Mairan in Paris, the other to Hell in Vienna, and the third to Lambert in Berlin. Some other explanations, suggested in the nineteenth century, will be discussed in Chapter 6.

4.1 Mairan's explaining away

Without saying so explicitly, Mairan tended to believe that the observations of Cassini and Short indicated the existence of a Venus moon, such as he had argued in the *Histoire* for 1741. He kept to this line of argument, which in 1762, in his autumn of life, led him to suggest a more detailed explanation of why the satellite had not been seen more often. "But what really to think about a satellite which is only seen two times in fourteen years and which thereafter is seen only fifty-four years later?" he asked. Apparently he was at the time unaware of the observation claims of 1761 to which he did not refer.[1] Although first and foremost a physicist, Mairan was also greatly interested in atmospheric and astronomical phenomena, such as the northern light and comets. These phenomena he sought to explain with his favourite, Cartesian-like hypothesis of a subtle material medium penetrating all bodies and parts of space. His writings on the aurora borealis culminated in 1754 with the massive *Traité physique et historiques de l'aurore boréale.*[2]

According to Mairan, the cause of the aurora borealis was the zodiacal light, the mysterious faint luminescence that Jean Cassini had discovered in 1683. Mairan suggested that the zodiacal light had its origin in the part of an extended, tenuous and varying solar atmosphere that reached as far as to the orbit of the Earth. As an interior planet, also Venus would experience an aurora, for the planet was in the range of the solar effluvia and, like the Earth, was rotating and covered by an atmosphere. This much was known, or thought to be known. Mairan's treatise was first published in 1732, as a supplement to the *Histoire et Mémoires*, and a substantial review of it appeared in the *Philosophical Transactions.* The reviewer, John Eames, a fellow of the Royal Society, summarized Mairan's view as follows:

> Are not the Inferior Planets, Mercury and Venus, almost always immersed in the Zodiacal Matter? and may not that be one Reason why 'tis so difficult to observe Spots in them? May not a Change, the Density, or Magnitude of the Solar Atmosphere, be one Reason why the Astronomers at Paris have not been able to observe those Spots in the Disk of Venus that have been taken Notice of, and described by Mr. Bianchini at Rome, a little before, since the Telescopes at Paris were of equal Length and Goodness?[3]

The hypothetical solar atmosphere might serve many purposes. One of them was Mairan's attempt to rescue Venus' moon, as he argued that under normal circumstances the solar atmosphere would make the moon invisible. He wrote as follows:

[1]Mairan 1764, p. 164. An almost identical version of the memoir appeared in *Journal des Sçavans*, August 1762, pp. 528–533. Given that Mairan read his first memoir to the Academy on 8 May 1762, about a year after Baudouin gave his addresses on the satellite of Venus, his silence with respect to Baudouin and Montaigne is puzzling. It is hard to believe that this work was unknown to him. Mairan's defense of the satellite of Venus was briefly mentioned by the Greifswald astronomer Lampert Röhl in a book of 1768 dealing with the Venus transits (Röhl 1768, p. 141).

[2]This was the second, much enlargened edition. The first edition appeared in 1732. On Mairan and his theory of the aurora borealis, see Kleinbaum 1973, pp. 203–228, who notes the connection to the Venus satellite on p. 225. See also Briggs 1967.

[3]Eames 1735, p. 256. .

> Venus' satellite and its principal planet are almost always submerged in the Sun's atmosphere, as we can show from the position and dimensions of this atmosphere; the satellite is thus almost always enveloped by a fluid matter more or less dense, which hides it totally or partly from us, and it is complicated by its smallness and by the structure of its little reflecting surface; I think it is to this variable cause we must attribute its fortuitous appearance and its long disappearances, whereas we always see its planet rather luminous both because of its size, forty or fifty times greater, and because of the structure of its reflecting surface.[4]

In this way Mairan attempted to explain the scarcity of observations of the Venus moon without abandoning them as spurious. He believed that under the right circumstances the moon would once again be observed and then vindicate the observations of Cassini and Short. The 84-year-old natural philosopher hoped "to request the astronomers not to reject a discovery upon which there is all reasons to believe that its object is very real, and which M. Cassini judged to be of a very great importance." Lalande, too, found it puzzling that so many astronomers, independently, at different locations and with different instruments, had observed the same phenomenon. If it was maintained that Montaigne had been seduced by optical illusions, it seemed necessary to assume that the other astronomers had fallen prey to the same illusions. According to Mairan and Lalande, this would be an unlikely coincidence. The observations had to be taken seriously, somehow to be explained.

In his article in the *Encyclopédie*, Lalande discussed various suggestions, admittedly of a somewhat speculative nature, in order to make sense of the observations. For example, the light from Venus' satellite might be much more feeble than the light from the satellites of Jupiter and Saturn, or perhaps the phases of Venus and its satellite, or the vaporous atmospheres supposedly covering them, might be responsible for the erratic observations.[5] Being among the period's most well known and influential astronomers, Lalande's words counted heavily in the astronomical community. Although he never supported the case of Venus' satellite unequivocally, for a period he was clearly sympathetic to the idea and did his best to give it a fair trial. His sympathy lasted until the early 1790s (see also section 3.1).

Lalande further referred to an earlier attempt of Giacomo Maraldi, the nephew of Jean Dominique Cassini, to explain the variability of a star in the

[4]Mairan 1764, pp. 164–165. See also the account in the *Histoire*, which according to Kleinbaum 1973, p. 242, was written by de Fouchy (Fouchy 1764).

[5]Encyclopédie II, p. 260.

constellation the Whale (Cetus), the Mira Ceti. First observed by the German astronomer David Fabricius in 1596, the phenomenon was discussed by the French natural philosopher Ismaël Boulliau in his *Ad astronomos monita duo* of 1667. Boulliau also offered an explanation of it, if not the correct one. The favoured explanation, adopted by Maraldi, was based on the assumption that the star was rotating and composed of two different parts, a bright and a dark region; if so, it might only be visible when the bright region turned towards the Earth.[6] Whatever the explanation, Mira had puzzled the astronomers because of its random character and the irregular periods between its observations. Perhaps, so Lalande suggested, the strange appearance and disappearance of the Venus satellite might be explained in a similar way. About twenty years later, the German amateur astronomer Friedrich von Hahn noticed in his observations of Mira Ceti a small variable star which only appeared randomly and which he therefore feared might be a deception, "like the supposed satellite of Venus."[7]

4.2 Ghost images?

In Vienna, the Hungarian-born Maximilian Hell (or Miksa Höll) reached a conclusion different from that of Mairan and Lalande, namely that the claims of a Venus moon were based on false images formed in the telescope and the eye. A Jesuit and priest, Hell was in 1756 appointed the first director of the observatory of the University of Vienna, founded the previous year.[8] Among his many duties was to edit the Vienna Ephemerides, which he did until his death in 1792. Hell kept a special interest in Venus and over the years he collected and analyzed the transit data in order to derive a more precise value of the solar parallax. He was involved in the 1761 transit project, reporting his results in the Vienna Ephemerides of 1762, and in connection with the transit of 1769 he was invited by the king of Denmark, Christian VII, to direct a small expedition to Vardø (or Wardhus) in the northernmost part of Norway. His expedition to Vardø, made in company with his Hungarian assistant Johann Sainovics, resulted in valuable, if also delayed and somewhat controversial data. Lalande intimated that the suspiciously delayed publication of the data implied that Hell had manipulated them, an accusation which

[6]Maraldi 1721, who also discussed other possibilities based on special properties of the luminous matter of which the starlight was assumed to consist. The brightness of the star was known to vary with a period of about 332 days, but the variation was not regular. For a survey of Mira Ceti, see Joy 1959. For the continual fascination of Mira Ceti and dark objects in the universe, see Jean-Sylvain Bailly's (1736–1793) wide-ranging chapter on dark and luminous celestial bodies in Bailly 1779, vol. 2, pp. 681–732.

[7]Hahn 1801, p. 197.

[8]On Hell's life and career, see Sarton 1944.

was widely accepted for more than a century. However, this is a subject of no direct relevance to the question of the satellite of Venus.[9]

Figure 4.1: Maximilian Hell dressed in a kind of Lapp costume during his stay at Vardø in 1769 to observe the Venus transit. Engraving from his *Observatio transitus Veneris ante discum Solis* (Copenhagen, 1770).

In 1757–58 Hell had observed Venus with both a refractor and a reflector and seen an ill-defined object near the planet. When he slowly moved his eye towards the eyepiece he found that the object changed into a moon-like image of Venus and with the same phase. But when he moved his eye perpendicular to the tube, the image (but not the planet) moved in the same way. Referring to the year of 1757, he wrote: "These trials, often repeated, went on for about an hour; and for that reason I concluded that I had been fooled by some optical image of false light through the Gregorian telescope, brought to the lens, perhaps, by the reflection in some polished and glittering part of the telescope's interior; and so I judged the observation to be an illusion which did not merit insertion in the diary of

[9] In addition to Sarton 1944, see also Woolf 1959, pp. 176–179 and Nielsen 1957b (in Danish). Hell's late rehabilitation was due to Simon Newcomb's (1835–1909) careful detective work. See Newcomb 1883.

observations." The next year he continued his optical experiments:

> I moved, extremely carefully and with the uttermost slowness of motion, the ocular nearer to the aperture of the telescope until I saw the right image very distinctly; and how strong was my wonder when I saw that this spurious light changed into a satellite of Venus, having the same phase as Venus itself! See there! (I said to myself), *this phenomenon is very similar to the one that M. Cassini saw earlier in Paris and M. Short recently in England.*[10]

In his Ephemerides for 1761 he encouraged astronomers to look for the Venus moon in connection with the transit, such as he did himself, albeit with no success. By the summer of 1761 at the latest, he had come to disbelieve in the reports of a Venus moon, such as he told of in a letter to Lacaille.[11]

Having conducted his series of careful experiments, Hell reached the conclusion that the image was formed first by the convexity of the cornea and then, by a second reflection, from the concave face of the meniscus lens or, in his case, the eyeglass.[12] He further found that the magnitude of the image would depend on the proportion of the curvature of the reflecting surfaces. Under certain conditions he found that he could always produce a spurious satellite, not only of Venus but also of Mars or Jupiter. As indicated above, these conditions included a very accurate position of the eye and a special, very delicate motion of it. According to the Viennese astronomer, this explained the few and erratic sightings of the spurious Venus satellite: Most observers, unaware of these special conditions, had been unable to recover the image they had once accidentally seen. Hell published an elaborate essay, *De satellite Veneris*, on these and other findings in 1765 and he also included it in his Vienna Ephemerides of the following year. Like other of his works, it was written in Latin. In this essay he included a careful review of all the observations of the supposed satellite of Venus known to him.[13] He summarized his critical essay in a number of rhetorical questions:

> Why would it [the image of the Venus moon] only appear to M. Montaigne at Limoges in 1761, and to M. Raedkier in Copenhagen 1764, and the same year to M. Montbarron in Auxerre, and not to others?

[10] Hell 1765, pp. 29–30.

[11] On the "secret letter," which somehow came to be known by Montaigne and to which he replied, see Hell 1765, p. 6, Lambert 1775, p. 180, and Schorr 1875, p. 71.

[12] The history of astronomy is rich of examples of observations based on false images. For a particularly interesting case, which in some respects illustrates Hell's hypothesis, see Sheehan 1988, pp. 204–208, and Sheehan and Dobbins 2003.

[13] Hell 1765 and Hell 1766. For later critique of his conclusions, see Schorr 1875, Webb 1876 and Stroobant 1887a. *De satellite Veneris* was also printed in *Nova acta Eruditorum*, February–March 1768, pp. 49–126.

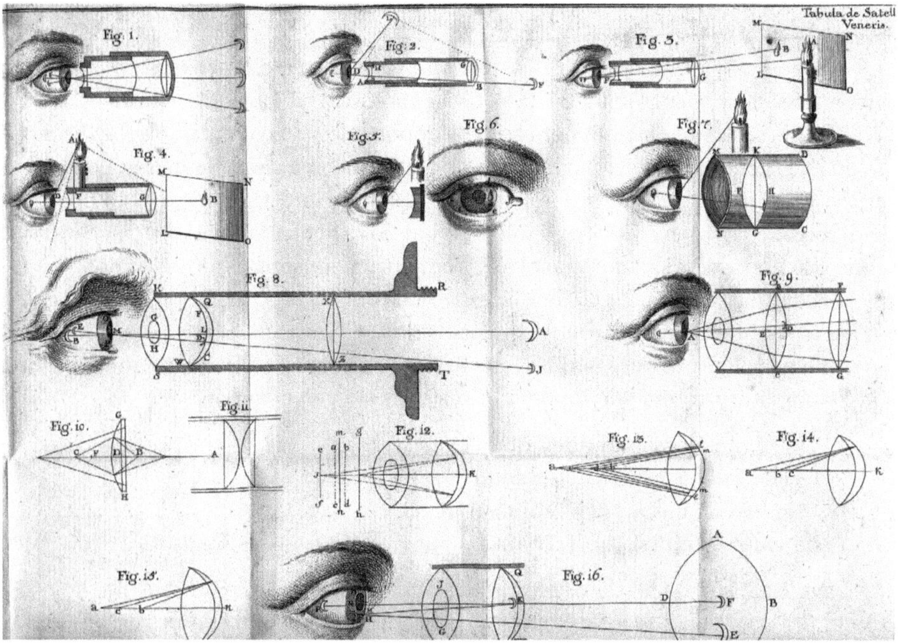

Figure 4.2: In his *De satellite Veneris*, Maximilian Hell made a series of optical experiments in order to account for the observation claims of a satellite of Venus. He concluded that the image of the strongly luminous planet was reflected both in the lenses and in the eye's cornea, and that this gave the impression of a satellite with the same phase as that of the planet. He further concluded that the image of a satellite only occurred if the eye was held in a special position relative to the eyepiece of the telescope tube. His argument was widely accepted and repeated in the literature of the late eighteenth century and the nineteenth century.

> How is it that during the same year of 1764, throughout the months of March, April, May and June I saw it very often, perhaps on the same days that M. Messier in Paris sought for it in vain, while they saw it in Copenhagen and Auxerre, but could not see it again at the same locations? Why, then, did it appear to me as often as I wanted, and to others not? Was that because I had clear skies while they all the time had clouds? To what cause could they [the protagonists of the satellite] ascribe it that all these times I saw it through two Gregorian telescopes, but never through the two much better Newtonian telescopes, although

I tried it often myself and offered others to try it?[14]

Together with other of his astronomical works, Hell's treatise on the false moon of Venus was translated into German shortly before his death in 1792.[15] The Viennese Jesuit astronomer used the opportunity to comment on the discovery claim of Scheuten of which he had not originally been aware. He had no faith at all in the observations made in Crefeld and was surprised that Lambert apparently had considered them to be reliable. "What Mr. Scheuten saw on the Sun on 6 June 1761 between 12 and 13 o'clock, apart from sunspots, I do not know; he has most definitely not seen the satellite of Venus."[16] The translator of *De satellite Veneris* was Longinus Anton Jungnitz, a student of Hell who since 1789 served as professor of astronomy and physics at the University of Breslau. Jungnitz used the opportunity to include a poem of his own on Venus' moon:

Wie kam es, dass man bey der Venus nichts
Den himmlichen Trabanten fand?
O sie bedarf ihn an dem Himmel nicht,
Die ganze Welt ist ihr Trabant.[17]

An English translation may go like this: "How did it come that the heavenly satellite / was not found at Venus? / O, it did not need to be in the heavens, / for the entire world is its satellite."

Apparently independent of Hell's ingenious hypothesis, his fellow-Jesuit Roger Boscovich came in 1767 to a somewhat similar conclusion.[18] Having stayed for a year in Paris 1759–60, Boscovich had good connections to influential French scientists. As early as 1748 he had become a correspondent of the Académie des Sciences, and while in Paris he formed close links to Mairan, Lalande and Alexis Claude Clairaut in particular.[19] As he was respected in French scientific circles, so was it the case in British scientific circles. Like so many contemporary astronomers, he was engaged in the preparations for the 1761 Venus transit, on which subject he contributed an article to the *Philosophical Transactions* (shortly thereafter, he was elected a foreign member of the Royal Society).[20] His plan was to observe the transit from Istanbul, but he arrived too late to witness the event. The Royal

[14] Hell 1765, p. 89.

[15] Hell 1792, which was the second volume of a four-volume work published 1791–94. The editor and translator, L. A. Jungnitz (1764–1831), published on electricity, meteorology and lunar eclipses.

[16] Hell 1792, p. 113.

[17] Ibid., p. 120.

[18] Boscovich 1767. This work was translated into German the same year by the Viennese Jesuit and priest Karl Scherffer (1716–1783).

[19] On Boscovich and the Paris Academy, including his strained relations to d'Alembert, see Pappas 1996. Issue 4, volume 49, of *Revue d'Histoire des Sciences* is a special issue on Boscovich.

[20] Boscovich 1761. He read the paper to the Royal Society on 19 June 1760.

Society later invited him to lead a voyage to California to observe the 1769 transit, but this plan was cancelled for political reasons.

As mentioned, Boscovich was acquainted with and for a while a colleague of Lagrange, the astronomer who in early 1761 believed to have observed the satellite of Venus. In an optical treatise of 1767, the *Dissertationes quinque ad dioptricam pertinentes*, he included an addendum on "The secondary images seen through lenses, and the apparent satellite of Venus" in which he explained the sightings of the Venus satellite as optical illusions caused by reflections in the ocular lens of the telescope as well as in the eye's cornea. His explanation was thus of the same kind as the one proposed by Hell. Strangely, Boscovich did not mention Hell at all or his work of the previous year, although it must have been known to him (the works of Hell and Boscovich were published by the same Viennese printer). In fact, there is no indication in the literature that Hell and Boscovich were in contact with one another.[21]

Figure 4.3: The erudite Jesuit scientist Roger Boscovich, here shown on a Yugoslavian note from 1981, did important work in natural philosophy, optics and astronomy. He was active in the preparations for the 1761 Venus transit observations. Familiar with the claims of a Venus moon, he argued that the sightings were caused by optical illusions.

[21] Between 1760 and 1786, Boscovich wrote more than 200 letters, but none of them was addressed to Hell. See Mandrino, Tagliaferri and Tucci 1986.

Boscovich did not say explicitly that all the satellite observations were due to reflections, nor did he deal with the observations in nearly the same detail as Hell. Nonetheless, the message of his essay was clear: he had no faith in the satellite of Venus. He explained what might have fooled the observers in the following way:

> The part of the light that is reflected at the ocular lens can be reflected once more by that lens towards the position of the pupil, and end in the bottom of the eye at a place different from the one to which the direct rays proceed to form the primary image; and these [the secondary rays] can at that point form a secondary image and exhibit something like a satellite. ... This can also happen by two reflections of which the one is in the cornea and the other in the ocular. ... The second reflection in that case [after the reflection in the cornea] can occur both in the first and in the second surface of the ocular glass.[22]

According to Boscovich, the effects would only arise if the curvatures of the ocular and the cornea were of the right order. Moreover, "It is not sufficient that the rays reflected from the ocular arrive at the pupil, but they also have to be sensibly parallel, for which some particular curvature of the ocular lens is necessary, to be determined by the curvature of the cornea."[23]

Sightings of "ghost" images were common and well known to astronomers, but not of quite the type that Hell and Boscovich described. In 1761, the prominent Swedish astronomer Per Wilhelm Wargentin, perpetual secretary of the Swedish Academy of Science, thought for a brief while that he had seen the Venus moon; but he soon discovered that the "moon" turned when he rotated the telescope tube axially. Wargentin had a high reputation among contemporary astronomers, and his experience was sometimes quoted in support of Hell's ghost-image hypothesis. For example, this is what Lalande did in his textbook *Astronomie*.[24] On 1 June 1777, when Venus was almost at conjunction at the upper limb of the Sun, the Swedish astronomer tried to observe the supposed moon which according to Lambert's prediction should be visible on this particular date. But he saw nothing. "I have always doubted if Venus were accompanied by a satellite," he wrote in the proceedings of the Swedish Academy of Sciences. "I am convinced that the astronomers, who believe to have seen it, have been the victims of optical delusion."[25]

---

[22] Boscovich 1767, p. 287. The translations from Hell and Boscovich have kindly been made by Henk Bos.

[23] Ibid.

[24] Wargentin 1780; Lalande 1792, vol. 3, p. 211.

[25] Wargentin 1780. On Wargentin and his observations of Venus, see the detailed biography Nordenmark 1939, especially pp. 174–197 (in Swedish but with an extensive French summary). Wargentin's Venus observations of 1 June 1777 were in response to the prediction made by

Johann Elert Bode reported a similar experience in his *Astronomisches Jahrbuch* for 1792, in which he referred to an observation he had made in 1788 in Berlin: "On the 21st of June I saw in a 3.5-foot Dollond telescope a mirror image of Venus originating in the eyepiece; at a certain increased distance of the eyepiece from the objective lens, Venus itself became unclear whereas the image became clearly visible, and by rotating the eyepiece it turned around Venus."[26]

On the other hand, if the observations of Wargentin and Bode were taken to count in favour of Hell's hypothesis, so the observations of Mayer in 1759 and Horrebow in 1764 would have to be taken as evidence against it. Both astronomers noted specifically that the bright spot did not move when they changed the axial direction of the telescope. Moreover, it was difficult to make Hell's explanation agree with Short's observation of 1740. Short had observed the supposed satellite for an entire hour, using different magnifications, and during this time he must frequently have moved the position of his eye. As Lambert was to point out, it was hard to imagine that the conditions of Short's observation was as fine-tuned as required by Hell's hypothesis.[27]

4.3 Lambert's orbital elements

The polymath Johann Heinrich Lambert was not foreign to the idea that there were – or once had been – unobserved objects in the solar system. In the first letter of his *Cosmologische Briefe über die Einrichtung des Weltbaus*, published 1761, he suggested that there might have been a planet between Mars and Jupiter, a suggestion that may be seen as an anticipation of what eventually became known as the Titius–Bode law: "And who knows whether already planets are missing which have departed from the vast space between Mars and Jupiter?"[28] He was not at that time acquainted with the few observations of the Venus moon, and he stated unequivocally that the solar system comprised ten satellites in all. At the time he published the *Cosmologische Briefe*, Lambert led a restless life in Germany and Switzerland. His luck improved a few years later, when he became a member of Frederick the Great's Academy of Sciences in Berlin, a position which started in early 1765 and lasted until his death twelve years later.

The year of publication of the *Cosmologische Briefe* was also the year of the Venus transit. In the company of some friends, Lambert observed Venus entering the Sun's disk on 6 June 1761. At a time during the transit, which he witnessed in

Lambert, to be mentioned in section 4.3.

[26] *Astronomisches Jahrbuch* (for the year 1792), published 1789, p. 254.

[27] Lambert 1775, p. 179.

[28] Lambert 1976, p. 57.

Figure 4.4: The German-Swiss polymath Johann Heinrich Lambert contributed to a variety of fields, including physics, mathematics, astronomy and philosophy. In the 1770s he became interested in the problem of Venus' satellite and derived from existing observations the orbital elements of the satellite. It is unclear if he believed in its real existence.

Augsburg, the Sun became covered by a thin veil of clouds, which made it possible to observe it with the naked eye. Lambert recalled:

> Not only did some of the spectators see Venus, they said that they had also seen an even smaller [object]. At that time I was unacquainted with the satellite, and for this reason I merely replied that the smaller Venus must be a sunspot, as such often occur. I did not take a closer look, for I had seen plenty of sunspots and did not know that there still was something to search for and observe. To be sure, now I would have wished that I had paid more attention to it.[29]

Whatever Lambert's early acquaintance with the problem of Venus' satellite, the idea soon fell in disrepute. He seems not to have taken any interest in the question during the 1760s. But in 1774-76 he decided to reconsider the problem in detail, which he did in a memoir of the Berlin Academy and also in two articles in the *Astronomisches Jahrbuch oder Ephemeriden*, the important publication

[29]Lambert 1776, p. 188.

series that Lambert established.[30] Since its beginning in 1774, the yearbook was published by the Berlin Academy of Sciences of which Lambert, as mentioned, had become a member in 1765. The more comprehensive article in the *Nouveaux Memoires* of the Berlin Academy, written in the summer of 1774, was reviewed in the *Allgemeine deutsche Bibliothek* by Abraham Gotthelf Kästner, professor of mathematics in Göttingen and a friend and correspondent of Lambert.[31] Kästner refrained from offering his own view on the controversial question of the existence of the Venus moon and sufficed to give a neutral account of Lambert's work. However, he found the subject to be fascinating, such as witnessed by a poem he wrote on *Der Venustrabant.* In its original German it goes as follows:

Das ihr Adonis noch am Himmel um sie geht,
Von Wahlen ward zuerst Cythere so geschmäht.
Sie haben freilich stets die Weiber im Verdacht;
Manch Sternrohr hat umsonst den Cisisbee bewacht.
Zu zeigen hat sich ihn einst Lambert unterstanden,
Und die Verläumdung ward zu Schanden.
So ist's am Himmel nur; man sieht Trabantenheere
Auf Erden leicht um jede Cythere.[32]

A rough English translation goes as follows: "That your Adonis still turns around her in the heavens, / this choice Cytherea at first found disgraceful. / Of course they always hold the women in suspicion. / Many a telescope has watched over a cicisbeo in vain. / Lambert once ventured to show his [the satellite's] presence, / and the calumny became a disgrace. / Only in the heavens is it like that; armies of satellites are / easily seen around every Cytherea on Earth."

Leonhard Euler, at the time residing in St. Petersburg at the Imperial Academy, was not pleased with the memoir from Berlin. In a letter to Lagrange, the great mathematical physicist, he commented on the reputed satellite of Venus. Without mentioning Lambert by name, he wrote: "Perusing the latest volume of the Berlin memoires I was not surprised at all that he [Lambert] raises yet

[30] Lambert 1773, Lambert 1775, and Lambert 1776. The Berlin *Astronomisches Jahrbuch* was from 1777 to 1829 edited by Bode, with whose name the yearbook is usually associated. On Lambert's occupation with the moon of Venus, and the connection to his cosmological world view, see Blumenberg 1975, pp. 609–626. Blumenberg argues that the two works represent the same "thought style."

[31] *Allgemeine deutsche Bibliothek* **27** (1775), 84–85. Kästner (1719–1800) is today best known for his work on parallel theory which inspired Lambert's important research on the topic and indirectly influenced the foundation of non-Euclidean geometry in the early nineteenth century.

[32] Kästner 1841, pp. 79–80. Apart from his works as a scientist, Kästner was also known as a prolific writer of occasional poetry. His poem on Venus' satellite, written about 1778, was one of many hundred *Sinnedichte.* The term *cicisbeo* (or cicisbee), derived from Italian, was widely used in the eighteenth century for an escort or lover of a married woman.

another question of Venus' satellite; and that even of the kind that turns upside down all the principles of astronomy. I will never come to believe that the principle of sufficient reason can be used so boldly as a theatrical effect."[33] It is unknown what caused Lambert to take up the subject, which at the time was not considered very interesting by the astronomers. It may have been a letter he received from Hell, in which the Viennese astronomer denied the existence of the satellite, a conclusion Lambert was not prepared to accept without further investigation.[34]

Aware of Baudouin's earlier determination of the orbit of the Venus satellite, Lambert set out to do a better job. After all, contrary to Baudouin he was an accomplished mathematician and astronomer. There is no indication that he sought to consult either Baudouin or other of the actors involved in the observations of Venus' moon. Lambert did not commit himself to either the existence or non-existence of the satellite, but reasoned that *if* the satellite were real it must be possible to calculate its orbital elements. He based his calculations not only on Montaigne's data but also took into account the observation that Scheuten had reported to him as well as the work of Roedkier and Horrebow in Copenhagen. For the eccentricity of the satellite's orbit, he found $e = 0.195$, a little less than that of Mercury, and he derived the orbital inclination to be 64 degrees. For the period of revolution he arrived at 11 days and 5 hours. Moreover: "I find ... that if the diameter of the Earth is = 1, the diameter of Venus must be = 0.97, of the satellite = 0.28, and of the Moon = 0.27. It further follows that the average distance of the Venus satellite is as large as $64\frac{1}{2}$ times the radius of the Earth, or $66\frac{1}{2}$ times the radius of Venus."[35] His data were thus in rough agreement with those previously obtained by Baudouin on the basis of Montaigne's measurements.

From the associated tables of the motion of the satellite, Lambert found that it would not have been visible towards the Sun in either 1761 or 1769, thus offering an explanation of the failure to observe it during the transits. During the first transit it would have passed just below the solar disk, and during the second above it. On the other hand, he calculated that on 1 June 1777 it should be possible to see the satellite in front of the Sun, close to its centre: "Venus will then pass 15 minutes above the upper limb of the Sun. When the satellite is in the lower part of its orbit it is to be expected that it will pass in front of the Sun. And should this really be seen, it will be an important contribution to decide the question if Venus has a moon."[36] This was a genuine prediction, but not a fortunate one.

[33]Letter of 23 March 1775, in Euler 1862, pp. 586–587. Euler (1707–1783) left Berlin for St. Petersburg in 1766, when he was replaced by Lagrange (1736–1813), who stayed in Berlin until 1787.

[34]On Hell's letter to Lambert, see Bopp 1915, p. 74.

[35]Lambert 1775, p. 183.

[36]Lambert 1776, p. 191.

When the date appeared, no satellite was seen by the astronomers who looked for it in Berlin, Vienna, Paris, Stockholm and Copenhagen. Johann Staudacher, an amateur astronomer in Nuremberg, searched for it on the date, but "I did not see the Venus satellite, even though I looked for it until 4 o'clock."[37]

It is unknown how Lambert responded to the failure, which may not have come as a complete surprise to him. He knew about the objections against the satellite of Venus, not only from the published sources but also from a letter he received from Hell in early 1777. Although praising Lambert as an eminent mathematician and theoretical astronomer, Hell disagreed completely with him with regard to the question of Venus' moon and the observational evidence in favour of it.[38] Lambert died prematurely a few months later, on 25 September 1777, and seems not to have commented on the unfortunate prediction.

In his two papers in the *Astronomisches Jahrbuch* Lambert did not comment on the mass of Venus, but this he did in his earlier and more extensive contribution to the Berlin memoir. He used the formula (*) given above in section 3.1, only with data somewhat different from those used by Baudouin, namely: $a_s = 64.5$ Earth radii, $a_m = 60.25$ Earth radii, $T_s = 11.2175$ days and $T_m = 27.3215$ days. Inserting the ratios $a_s/a_m = 1.07$ and $T_m/Ts = 2.43$ in the equation, he got $M_V/M_E \cong 7.278$. Contrary to Baudouin, Lambert adopted the standard value $R_V/R_E \cong 0.97$ and from this he found the relative density of Venus to be $\rho_V/\rho_E \cong 8$. It may seem incredible that Venus, the sister planet of the Earth, is so much more massive, but Lambert sought to justify his result by relating it to the planet's power of light reflection:

> A planet like Venus which reflects so much light does not seem to be a very porous body. We know that the density of Jupiter is only 1/5 and that Saturn's is only 1/10 of that of the Earth. This could let one to believe that the density of the planets increases gradually as they come closer to the Sun. Even so, however, the density of the Sun itself is only a quarter of that of the Earth. So, there are no means to conclude much from all this.[39]

Perhaps he did not find his reasoning convincing by second thought, which may have caused him to omit Venus' mass altogether from his later articles in the *Astronomisches Jahrbuch*. It is unclear if Lambert believed in the existence of the Venus satellite, but in his treatise of 1773 he expressed himself as if he did. He ended his memoir with reflections on the use and purpose of the supposed satellite.

[37] Quoted in Wolf 1857, p. 276.

[38] The letter, dated 15 February 1777, is printed in Hell 1792, pp. 114–120.

[39] Lambert 1773, pp. 244–245. The idea that "the planets that are nearer to the Sun are also denser" can be found in *Principia*. Newton 1999, p. 814, related the density of the planets to the amount of heat they received from the Sun.

Like most other natural philosophers in the period, he was convinced that Venus and the other planets were inhabited:

> This satellite being so little visible, its eclipses in the shadow of Venus will never be of great use for us. Furthermore, they last not for long, and the great inclination of its orbit makes them happen only very rarely. On the other hand, the shadow which the satellite casts on Venus is an extremely faint penumbra; the inhabitants of Venus see their satellite under an angle of $14\frac{1}{2}$ minutes, whereas the Sun is seen under an angle of 44 minutes. Because of this, daylight is only reduced by 1/10, which is nothing for light such as the one that comes from the Sun. Thus, this satellite not only hides itself from our view, but it conceals moreover the traces of its movements.[40]

More than a century later the Belgian astronomer Paul Stroobant (of whom more in section 6.2) argued that the values obtained by Lambert were inconsistent and, therefore, "the existence of a true satellite is definitely ruled out."[41] Stroobant did not use the systems considered by Baudouin and Lambert, but the Venus-Sun and the satellite-Venus systems, that is, he adopted the same approach that Lalande had followed in 1781. By inserting in equation (**) the values $T_V = 224$ days 7 hours, $T_s = 11$ days 2 hours, $a_s = 63$ Earth radii, and $a_V = 16846$ Earth radii, he arrived at the result

$$\frac{M_V}{M_\otimes} = \frac{1}{42336}.$$

Comparing it with the modern ratio, based on perturbation theory, namely

$$\frac{M_V}{M_\otimes} = \frac{1}{412150}$$

Stroobant found that the mass determined from Lambert's data was 9.74 times too great and concluded that he had refuted the hypothesis of a Venus satellite. He had been foreshadowed by the English astronomer John Russell Hind, who in a much earlier comment on Lambert's analysis concluded that there was one "fatal objection" to it, namely: "If it were correct, the mass of Venus would be ten times greater than the value found from theory by other methods."[42] Contrary to

[40] Lambert 1773, p. 248. On Lambert and pluralism, see Crowe 1999, pp. 55–59. In his *Cosmologische Briefe*, Lambert not only made all heavenly bodies (comets included) inhabitable, he also believed – in accordance with the principle of plenitude – that there were on each of these "innumerable inhabitants of all possible kind and form" (Lambert 1976, p. 82).

[41] Stroobant 1887a, p. 11. William Smyth, on the other hand, found Lambert's theory to be "very consistent" (Smyth 1844, vol. 1, p. 109).

[42] Hind 1852, p. 38.

Stroobant, Hind realized that the objection did not rule out the satellite of Venus, but only questioned the Baudouin-Lambert version of it.

The criticism of Hind and Stroobant carried no historical force. In 1773, when Lambert wrote his essay, the perturbation method was not yet fully developed and had not been applied to the case of Venus.[43] Using perturbation theory, Laplace reported in the fourth edition of his *Système du monde* the value $M_V/M_E = 0.945$, which is rather far from the presently accepted ratio of 0.815.[44] Laplace's value is close to the one obtained by Baudouin, but the agreement in coincidental.

It may be worth summarizing the calculations made in the eighteenth century to find the mass of Venus relative to that of the Earth from the observed data of the planet's satellite. This is done in table 4.3, where the results are compared with the perturbation-based values obtained by Dionis du Séjour and Laplace and also the modern value.

Author	**Mass ratio, Venus/Earth**	**Method**
Baudouin 1761	0.98	Venus-satellite; Earth-Moon
Lambert 1773	7.28	Venus-satellite; Earth-Moon
Lalande 1781	8.65	Venus-satellite; Sun-Venus
Dionis du Séjour 1789	0.781	perturbations
Laplace 1813	0.945	perturbations
(modern)	0.815	perturbations

Table 4.1: The mass of Venus relative to the mass of the Earth. The first three rows presume the existence of a satellite of Venus.

[43] On the method of determining the mass and density of a planet by means of its satellite, as known in the late nineteenth century, see e.g. Dionis du Séjour 1789, pp. 522–524, who for Venus reported $M_V/M_E = 0.781$ (p. 527). However, since he did not believe in the existence of a Venus moon, this value was based on estimates of the planet's perturbations on the orbit of the Earth.

[44] Laplace 1813, p. 217.

Chapter 5

A spurious but persistent satellite

Lambert's memoir on the satellite of Venus was for a long time the last serious study of the subject. By the 1780s the satellite was on its way out of astronomy, with the majority of astronomers either dismissing it or, more commonly, ignoring it. Lalande's mention of the subject in the *Dictionnaire de physique* of 1781 and later in his *Astronomie* of 1792, uncommitted but not clearly dismissive, was an exception. As we shall see, the rejection of the satellite of Venus did not mean that it was deemed to oblivion. This was far from the case. Still, from the point of view of the large majority of astronomers the question was no longer controversial. Whatever its mysteries (and these were many), it was agreed that Venus just could not boast of a moon.

5.1 Dismissed but not forgotten

A few years after Lambert's investigation of Venus, William Herschel engaged in a series of observations of the planet, a research programme he started in 1777. Herschel was particularly interested in the atmosphere and rotation of the planet, two phenomena he thought were connected. Venus' thick atmosphere had first been deduced by the great Russian chemist and polymath Mikhail Vasilyevich Lomonosov from observations made during the 1761 transit. Without knowing of Lomonosov's results, Herschel concluded that Venus had a dense atmosphere, and that its purpose was to shield the inhabitants from the excessive heat of the Sun. His belief in Venusian beings was by no means exceptional – Lomonosov, for one, believed in them too.

For several years Herschel looked carefully for the spots that might reveal an axial rotation, but was frustrated that in almost all of his observations they failed to turn up. The reason, he suggested, was the dense atmosphere that most often would hide the appearance of the solid surface. On 30 November 1789 he wrote in his observation diary: "No satellite visible. If she has one, it must be less in appearance than a star of the 8th or 9th magnitude; power 300."[1] This was all he had to say about the moon of Venus. Bode, another of the period's great astronomers, was no less sceptical. In a popular work on astronomy he referred briefly to the hypothetical satellite of Venus: "A few astronomers have sometimes thought to have seen a [Venus] moon, but, for various reasons, its existence is still most doubtful."[2] In a later work, published 1816, he repeated the judgment, this time mentioning the various observation claims from Fontana to Montbarron. "Since that time [1764] no real astronomer has boasted of having noticed a similar phenomenon at Venus."[3]

Among the specialists in observational planetary astronomy belonged the respected German amateur astronomer Johann Hieronymus Schröter, a contemporary of Herschel. In 1790 Schröter provided additional observational evidence for the Venus atmosphere, and he argued that he had seen exceedingly high mountains on the planet (as high as 40 km). He was also the first to find that the observed and theoretically expected phases of Venus do not agree exactly; and he noted that when Venus is close to the solar limb, it will appear to be surrounded by a thin and luminous ring. But his diligent observations of Venus, made from his observatory in Lilienthal near Bremen, never led him to entertain the idea of a satellite. In the *Aphroditographische Fragmente* of 1796, primarily a description of the 27-foot Lilienthal telescope, he exuberantly discussed the divine purpose and inhabitableness of the planets. Is Venus not a fully developed celestial body "because it probably has no companion, such as our Earth"? In a footnote he added:

> During my 15 years of observation I have never, in spite of all attention, found the slightest trace of either a real satellite or, in any telescope, a deceptive secondary image, such as the late Father Hell thought in his treatise of 1766.[4]

[1] Herschel 1912, vol. 1, p. 444. Originally published as "Observations on the planet Venus," *Philosophical Transactions* **83** (1793), 201–219.

[2] Bode 1807, p. 35. Calling the existence of the moon "still most doubtful" (*noch sehr zweifelhaft*) may suggest that Bode was not yet willing to write off the satellite.

[3] Bode 1816, p. 357.

[4] Schröter 1796, p. 193. A modern selection of Schröter's important observations of Venus, including papers by Herschel and Harding, appears as Schröter 1995. For an appreciation of Schröter's astronomical work, see Moore 1960.

Satellite or not, Schröter was firmly convinced that Venus was inhabited with living beings adapted to the planet's particular physical environment. In an English translation appearing in the *Philosophical Transactions*, he said that nature had raised on Venus

> ... such great inequalities, and mountains of such enormous height, as to exceed 4, 5, and even 6 times the perpendicular elevation of Chimboraco, the highest of our mountains. Thus are we, by these observations, led to a farther contemplation of the immense, and yet analogical variety with which the great Author of nature has dignified his works, as well in the greater objects, as in the smallest microscopic atoms; and the incessant novelty of combinations with which he has adorned them.[5]

Although the moon of Venus was no longer seriously considered by the majority of astronomers, it was far from forgotten. Jean-Sylvain Bailly, the French astronomer and pioneer historian of astronomy, included the subject in volume 2 of his *Histoire de l'astronomie moderne*, and it was also covered in some detail in Montucla's *Histoire mathématiques.*[6] Bailly's attitude was sceptical, but not dismissive, and he recommended to "suspend our judgment: its existence cannot be affirmed, nor can it be denied." As far as physical and philosophical arguments were concerned, and especially those based on final causes, he found them to be unsatisfactory as they could be used as an argument both for and against the existence of the satellite. The question could only be resolved by means of observations, and so far these had been disappointing. Montucla agreed, but was more critical in his evaluation of the observation claims, which he considered to be of no real value. Like most other commentators, he found Hell's explanation to be credible.

The alleged satellite of Venus not only appeared in the scientific literature, it also appeared in some of the encyclopedic works from the turn of the century. Johann Samuel Gehler's widely read *Physikalisches Wörterbuch* included in its article on satellites an account of the observations of the Venus moon, starting with Fontana and ending with Lambert. The author of the article expressed support of Hell's explanation of the observations in terms of ghost images and concluded

[5]Schröter 1792, p. 337. Venus does have high mountains – the highest peak has an elevation of 11,580 km – but that became known only much later. Chimboraco, now Chimborazo, a defunct volcano in Ecuador, has a height of 6,310 m. The greater height of Mount Everest and other Himalayan mountains was only established in the mid-nineteenth century. On the history of the high mountains of Venus, see Baum 1973, pp. 51–69.

[6]Montucla 1802, pp. 15–16; Bailly 1779, vol. 2, pp. 407–411. Most of Montucla's volume 4, which dealt with the astronomical sciences and appeared only posthumously, was written by Lalande, a lifelong friend of Montucla. It is uncertain if the section on Venus' moon was written by Montucla or Lalande.

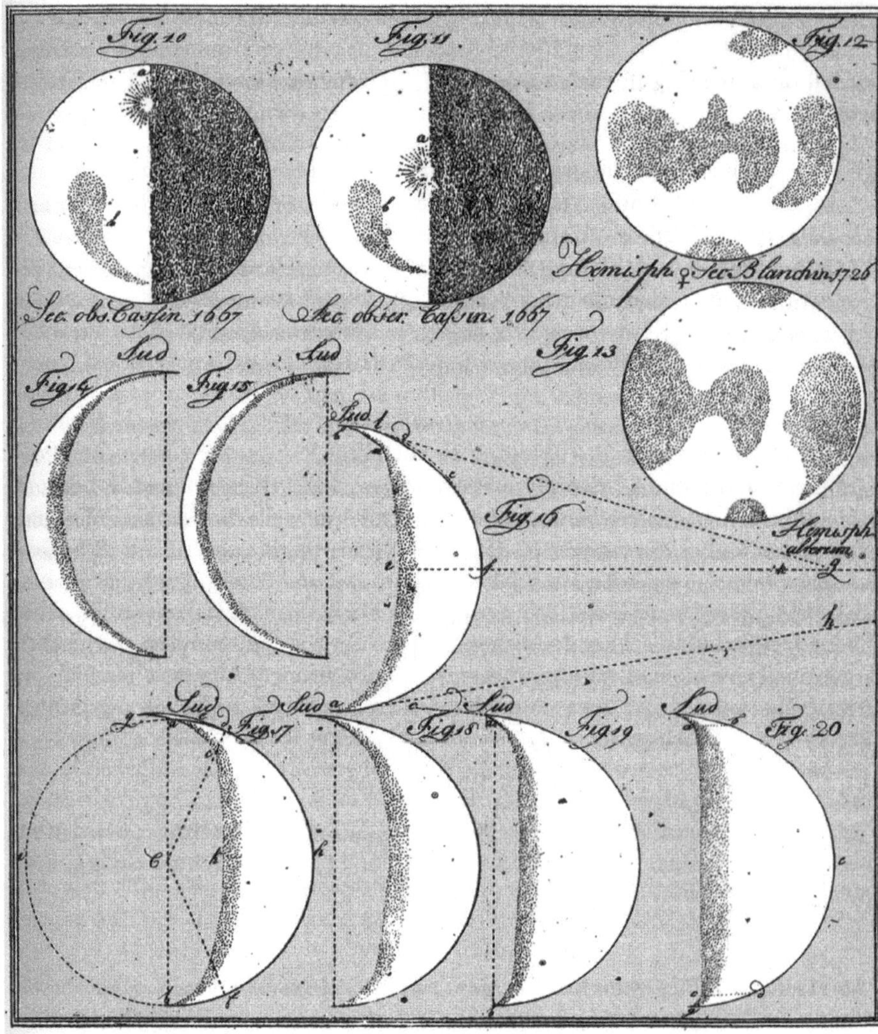

Figure 5.1: The German astronomer Johann Schröter made careful observations of the shape and surface of Venus, but never saw a satellite following the planet. His drawings of Venus included several details that do not exist, such as shown in this plate from his *Beobachtungen über die sehr beträchtlichen Gebirge und Rotation der Venus* (Schröter 1995, p. 23).

that the observation reports "have been due to errors on behalf of the observers."[7] Charles Hutton, a distinguished applied mathematician and fellow of the Royal Society, was more positively inclined in his *Mathematical and Philosophical Dictionary* of 1795. In his account of the satellite of Venus he emphasized the difficulties of observing a small moon belonging to an interior planet. "Indeed it must be acknowledged, that Venus may have a satellite, though it is difficult for us to see it."[8] Just like the *Encyclopédie* of 1778–82 included a detailed account of the satellite of Venus, so did its British counterpart, the *Encyclopædia Britannica* of 1801. Much in the style of the *Encyclopédie*, and more or less copying it, the British encyclopedia provided details about the observations of Cassini and Short, and also referred to the works of Montaigne and Baudouin.[9]

Venus' rumored companion found its way into the Göttingen professor Johann Erxleben's *Anfängsgründe der Naturlehre*, probably the most popular textbook in the physical sciences in Northern Europe. Edited by another Göttingen physicist, Georg Christoph Lichtenberg, the book was published in six editions between 1784 and 1794. After having recounted a brief history of the satellite of Venus, as usual starting with Fontana and ending with Lambert, Lichtenberg noted that "its existence is still very uncertain."[10]

The widespread knowledge of the Venus moon in the late eighteenth century may be further illustrated by a reference to Denmark, a country which scientifically was at the periphery of Europe. Esaias Fleischer, a civil servant and amateur scientist, wrote between 1786 and 1804 a huge work in 26 volumes, what he called a universal natural history. In the first volume of this great project he described the world system from the perspective of astronomy, if constantly with an eye on natural theology. Including a section on Venus' satellite, he referred to the standard chronology from Fontana to Roedkiær and Montbarron. Although Fleischer found it hard to accept Hell's reflection-image hypothesis, he admitted that the lack of confirmation during the transits in 1761 and 1767 amounted to hard evidence against the possibility of a satellite. For this reason he concluded that the existence of a Venus moon was "dubious."[11]

---

[7]Gehler 1798, p. 340. The second edition, published in 1833 and edited by the chemist Leopold Gmelin (1788–1853) and others, excluded the account of the former edition and simply stated in a footnote that it was unnecessary to deal with "the opinion, so certainly erroneous" that Venus possesses a moon (vol. 7, p. 64).

[8]Hutton 1795, vol. 2, p. 649.

[9]Most of the account on the satellite of Venus is quoted in Thornton 1804, pp. 329–332. The first edition of the *Encyclopædia Britannica* appeared in 1768–71.

[10]Erxleben 1787 (fourth edition), pp. 569–570. The Danish translation of the book, *Begyndelsesgrunde til Naturlæren* (1790), included references to the works of Roedkiær and Horrebow.

[11]Fleischer 1786, p. 909. Fleischer (1732–1804) was unaware of Scheuten's observation and of course also of the unpublished observations of his compatriots Roedkiær and Horrebow of 1761.

The article in the *Encyclopædia Britannica* and Bonnet's reference in *Contemplation de la nature*, mentioned above, furnished the main sources for Robert John Thornton who in 1804 advocated the controversial satellite hypothesis in a polemical reply to the chemist and mineralogist Arthur Aikin. Thornton, a London physician and naturalist, is best known for his magnificently illustrated tables *The Temple of Flora*, a work he produced in 1797–1810 and the title of which he borrowed from Erasmus Darwin, his hero.[12] Inspired by Darwin's cosmogonical ideas, he defended in the second part of his great work that Venus was followed by a satellite and that it was discovered in the eighteenth century. In a lengthy and extremely critical review in the *Annual Review*, the chemist and mineralogist Arthur Aikin took him to task for supporting such a ridiculous claim. How was it, he asked, that the satellite had eluded the sight "of Dr. Herschel, of the astronomer royal, and of every other astronomer of the world"? Surely it was nothing but a phantom:

> As we are told that the discovery was made in the last century, it must have been known to others as well as to himself; and as no trace of it is to be found in any professed treatise of astronomy, we may be allowed to doubt on the subject, and to suspect that Dr. Thornton has asserted what is not true.[13]

In his reply to Aikin's attempt to ridicule him, Thorncroft documented in detail that, whether or not the satellite really existed, he was in good company and better informed than his critic. The satellite could indeed be found in many professed treatises of astronomy.[14]

"The belief in a Venus moon has been abandoned, and for a long time there has been no talk about it."[15] So stated a popular astronomy book of 1828, undoubtedly correctly. Even French authors agreed that whatever the great Cassini had seen in his telescope, it was almost certainly not a moon of Venus. Lalande, who for a long time had been favourably inclined towards the satellite, came to agree with the majority view. In his *Histoire de l'astronomie moderne*, Jean-Baptiste Delambre dealt in great detail with Cassini's contributions to astronomy, including his observations of 1672 and 1686. But he briefly and rather superficially dismissed them in terms of the mechanism proposed by Hell. What Cassini had seen was nothing but "an optical illusion which is reproduced not only in the case

[12] Thornton 1804. The English naturalist R. J. Thornton (1768–1837) was a Darwinian *avant le mot*, and Erasmus Darwin (1731–1802) praised *The Temple of Flora* as having "no equal." On Thornton and Darwin, see Bush 1974.

[13] Aikin 1803, p. 880. The work under review was Thornton 1803. From 1803 to 1808 Aikin (1773–1854) served as editor of the *Annual Review*.

[14] Thornton 1804.

[15] Brückner 1828, vol. 2, p. 55.

of Venus, but also for the other planets and most stars."[16] Delambre obviously did not find the subject to be worth a closer examination.

Until the 1880s, scientific references to the satellite of Venus were few and scattered, although they were a standard ingredient in more popular works on astronomy. These typically told the story in the same way, relying on the same secondary sources and repeating the same errors and misunderstandings.[17] Most stories were historically framed , in the sense that they accounted for what earlier generations of astronomers had believed; and they also included a moral element, in so far that they highlighted the epistemic superiority of current knowledge over that of earlier periods. They never included unconditional knowledge claims that there was, in fact, a moon of Venus.

In so far as they considered the question at all, the majority of astronomers probably agreed with the verdict of the famous naturalist Alexander von Humboldt, who in his influential *Kosmos* wrote about the "so-called Venus moon" that it "belongs to the astronomical myths of an uncritical age."[18] Johann Mädler, director of the Dorpat Observatory and a leading figure in European astronomy, tended to agree, concluding that "neither Venus nor Mercury have satellites, and among the interior planets the Moon of the Earth is the only satellite." Mädler aired the possibility that the observations might be due to an unknown planet with about the same period of revolution as Venus, but only to dismiss it as an unlikely explanation. He preferred to support Hell's hypothesis of optical illusions:

> I have myself sometimes seen secondary pictures, not only at Venus but also at Jupiter and Saturn, but at every occasion I have been convinced that they were nothing but pure optical illusions. Because, when I brought the primary picture into the centre of the field of vision, then the secondary picture moved towards it from the other side and disappeared in the centre, merging with it.[19]

Later in the *Geschichte der Himmelskunde* he made the same point in a slightly different way: "I recall that once, in the beginning of my astronomical career, I saw Jupiter accompanied by five moons. Only after having turned the eyepiece several times and changed the focus to Jupiter was I satisfied that the fifth moon was nothing but an optical reflection of the primary planet."[20]

---

[16] Delambre 1821, p. 743.

[17] Examples are Anger 1862 and Tuxen 1861, pp. 215–216 (in Danish). We have not made any systematical study of the literature.

[18] Humboldt 1845–58, vol. 3, p. 539. Schorr 1875 protested against Humboldt's characterization (pp. v, x and 60).

[19] Mädler 1873, vol. 1, pp. 475–476. The idea of an unknown planet masquerading as the moon of Venus was developed by J.-C. Houzeau in 1884 (see section 6.1).

[20] Mädler 1873, vol. 2, p. 234.

5.2 Worthy of attention

Although the idea of a Venus moon was decidedly unpopular, not all astronomers were willing to discard it. Thomas John Hussey, an English clergyman and amateur astronomer, published in 1834 a review article in the *Astronomische Nachrichten* on the rotation of Venus. In this connection, he argued against the hypothesis of optical illusion as an explanation of the sightings of the Venus satellite. Hussey did not express belief in the satellite, but he did believe that it was "worthy of attention."[21] This cautiously positive view was shared by the influential French physicist and astronomer François Arago, permanent secretary of the École Polytechnique.

According to Arago, the enigmatic satellite was not to be easily dismissed. In his posthumously published *Astronomie populaire* he considered the subject and its history in a section entitled "What should be thought of the Venus moon?" Although not endorsing the discredited satellite, Arago would not reject it either. Perhaps, he suggested, the satellite or its surface was made up of a cloud-like material that strongly reduced its reflection of sunlight. He considered the reputed moon of Venus to belong to "the domain of the possible," but was careful not to elevate its status to the domain of the real.[22] A somewhat similar attitude was adopted by John Russell Hind, a respected astronomer and discoverer of minor planets. In his popular work *The Solar System* he included a section on the satellite of Venus and its observational history. He found it hard to accept that all observers from Cassini to Horrebow could have been mistaken. "It is a question of great interest," he concluded, "and must remain open for future decision."[23]

The British clergyman and amateur astronomer Thomas William Webb was one more mid-nineteenth-century astronomer who kept an interest in the satellite of Venus. In his *Celestial Objects*, a classic guide for amateur astronomy for more than half a century, he described it as "an astronomical enigma." It was, he wrote, "not easy to set aside the evidence of its occasional appearance." Although Webb admitted that some of the observations might be explained in terms of Hell's hypothesis of reflections, he denied that this was the case with Short's observation

[21] Hussey 1834.

[22] Arago 1854–60, vol. 2, pp. 538–542. The book appeared simultaneously in German, as *Populäre Astronomie*, ed. W. C. Hankel (Leipzig: O. Wigand, 1855–59), where the section on the Venus moon is in vol. 2, pp. 473–476. Although best known as a physicist, Arago (1786–1853) also contributed to astronomy and astrophysics. He was director of the Paris Observatory 1843–53 and during the 1842 solar eclipse he examined the Sun's chromosphere and determined that the limb of the Sun is gaseous.

[23] Hind 1852, p. 37. J. R. Hind (1823–1895) received the Royal Society's Gold Medal and the Lalande Medal. He discovered ten asteroids and was the first to observe the new planet Neptune from Britain. Obituary in *Monthly Notices of the Royal Astronomical Society* **56** (1896), 200–205.

of 1740.[24]

The cautious attitude of Arago, Hind and Webb was shared by Karl Theodor Anger, a mathematician and astronomer at Königsberg, who in 1862 argued that "the investigations of this matter should in no way be considered exhausted."[25] He realized that some of the observations might be due to optical ghost images but found it hard to believe that expert observers such as Cassini, Short, Montaigne and Horrebow had been fooled by their telescopes. And if this were really what had happened, why had the false images not turned up in later observations of Venus? This was the same objection that Horrebow had raised nearly a century earlier. The same point was also considered by Hermann Joseph Klein, a respected amateur astronomer who possessed a private observatory in Cologne and edited *Sirius*, a journal of popular astronomy founded in 1880. Klein leaned towards the hypothesis of ghost images but nonetheless presented the observations as "an astronomical enigma." He did not believe in the satellite of Venus, but, like several of his contemporaries, would not abandon it for good either.[26]

The only astronomer in the mid-nineteenth century who explicitly endorsed the controversial satellite of Venus may have been William H. Smyth, a British navy officer and distinguished amateur astronomer. In his classic compilation *A Cycle of Celestial Objects* he evaluated the controversial moon in a much more positive light than his contemporaries. "To the present moment it cannot be demonstrated that it is not in existence," he wrote, urging his fellow astronomers not to relinquish the search for it.[27] Smyth pointed out that the satellite, supposedly very small, would be extremely difficult to find. Because, when Venus is nearest to the Earth, and circumstances are most favourable for its detection, the dark side would be turned toward us. As to the generally held belief that the observations of the moon were due to ghost images or the like, he found it to be "dogmatic."

As we have seen, in the years about 1840 interest in the satellite of Venus was at a low point. Among the few works which referred to the subject, one in particular stands out because it made a connection to pluralism and natural theology. Thomas Dick, an Irish-born priest, teacher and author, was equally occupied with Christian education and natural science. A prolific writer, in several of his works he argued eloquently in favour of extraterrestrial life throughout the universe. Not only were the planets and their satellites inhabited, so were the Sun, the asteroids and Saturn's system of rings (only with respect to comets did he

[24]Webb 1859, pp. 45–46. According to Hockey 2007, p. 1200, T. W. Webb (1806–1885) was "the patron saint of British amateur astronomers." On Webb's life, see the obituary in *Monthly Notices of the Royal Astronomical Society* **46** (1886), 198–201.

[25]Anger 1862, p. 103. See also Lardner 1860, p. 206, which concluded that the observations of Venus' moon were probably due to illusions.

[26]Klein 1879, p. 118, which included a reproduction of Lambert 1775 and 1776.

[27]Smyth 1844, vol. 1, pp. 109–110.

express hesitation). To Dick and other pluralists the notion of habitable celestial bodies without life went counter to their Christian faith; for this faith told them that God had created matter for the sake of intelligent beings, supposed to be more or less comparable to humans. Relying on the old principle of plenitude they believed that if life could exist, it did exist. Although primarily a writer, Dick also made astronomical observations, some of which he published in the *Edinburgh New Philosophical Journal.* For example, in 1844 he reported observations of Venus at the time of its superior conjunction.[28]

Dick's *Celestial Scenery,* published in 1838, ran into six editions and about 25 reprintings. This successful work was primarily concerned with astronomy rather than theology, although in Dick's mind the two subjects were necessarily intertwined. What is of interest in the present context, the book included a detailed section on the satellite of Venus with quotations from Cassini and later observers. "There is a singular consistency in these observations, which it is difficult to account for if Venus have no satellite," he opined. Based on analogical reasoning he was inclined to believe that the satellite was real: "It is somewhat probable, reasoning *a priori,* that Venus, a planet nearly as large as the earth, and in its immediate neighbourhood, is accompanied by a secondary attendant."[29]

As to the primary planet, Dick was of course convinced that it was inhabited, for "the Creator has, doubtless, in this as well as in every other case, adapted the structure of the inhabitant to the nature of the habitation." He even calculated the number of Venusians, arriving at the astounding figure of 53.5 billions.[30] In another work, *The Sidereal Heavens* of 1840, he went even further and estimated the population of the entire universe to be about 6×10^{22} (one tenth of Avogadro's number!). Not only did he consider it "somewhat probable" that Venus was endowed with a moon, the same might also be the case for Mars: "If such a satellite exists, it is highly probable that it will revolve at the nearest possible distance from the planet, in order to afford it the greatest quantity of light. ... It is therefore possible, and not at all improbable, that Mars may have a satellite, although it has not yet been discovered." Moreover:

> It is no argument for the nonexistence of such a body that we have not yet seen it. ... The long duration of winter in the polar regions of Mars seems to require a moon to cheer them during the long absence of

[28] Astore 2001, p. 37. On Dick (1774–1857) as a pluralist, see Crowe 1999, pp. 195–202 and Hennesey 1999, pp. 62–65. Yeo 1986 provides an insightsful analysis of the British context of pluralism and natural theology.

[29] Dick 1838, p. 96 and p. 99.

[30] Ibid. p. 135. His population number for Mars was 15.5 billions, and for the Moon 4.2 billions; Saturn, including its system of rings, could boast of no less than 13,630 billion inhabitants. For comparison, the population of the Earth in 1838 was about 1.2 billion. Dick's table is reproduced in Crowe 1999, p. 199.

> the sun; and if there be none, the inhabitants of those regions must be in a far more dreary condition than the Laplanders and Greenlanders of our globe.[31]

Although Dick was sympathetic to the satellite of Venus, and found its possible existence to be congruent with both pluralism and natural theology, there never was a strong link between the two. As we have seen, Schröter described the architecture of Venus in the language of natural theology, but he did not believe that the planet had a companion.

Dick's interest in the satellite of Venus was noted by at least some astronomical writers. Hiram Mattison, an American Methodist minister and author of elementary textbooks of astronomy, was among the few who actually supported the hypothesis of a satellite. Referring to Dick, he found it "highly probable that such a body exists." Observations might still decide in favour of the satellite, he thought, "a question worthy of the attention of Lord Rosse, and the powers of his colossal reflector."[32]

David Brewster, the Scottish natural philosopher and eminent optician, was no less a pluralist than Dick. And yet he considered the observations of the supposed satellite of Venus to be nothing but deceptions. Relating to Wargentin, the Swedish astronomer mentioned in section 4.2, he wrote that "he had in his possession a good achromatic telescope, which always showed Venus with such a satellite, and ... the deception was discovered by turning the telescope about its axis."[33] Brewster was well aware of the history of Venus' moon, but he shared the standard view that the moon was not real. In *More Worlds than One*, a main work in the pluralist literature, he wrote: "In this group of [interior] planets no moon or satellite has yet been discovered, and it is probable that none exists. An atmosphere of great height, and of a peculiar constitution reflecting on the planet the light of the sun many hours after he has set, might in all of them supply the place of a moon."[34]

Nor did William Whewell, Master of Trinity College and a highly respected scientist and philosopher, endorse the hypothesis of a Venus moon. In his

[31]Dick 1838, p. 140.

[32]Mattison 1849 (fifth edition), p. 94. Apart from his lectures and books on astronomy, Mattison (1811–1868) was also known for his active support of the anti-slavery movements.

[33]Quoted in Dick 1838, p. 99 and also in Smyth 1844, p. 109. We have not been able to locate the original source. Neither Dick nor Smyth accepted Brewster's argument that astronomers in the eighteenth century had been deceived by optical illusions. As Dick pointed out, the satellite had been seen both with refractors and reflectors, and sometimes with different powers applied, for which reasons it was unlikely that the observations were caused by the instrument.

[34]Brewster 1854, p. 76. On Brewster's pluralism and his debate with the prominent scientist, philosopher and priest William Whewell (1794–1866) on this issue, see Brooke 1977 and Crowe 1999, pp. 313–351.

Astronomy and General Physics with Reference to Natural Theology of 1833, a work in the same tradition as Dick's, he argued that the satellites were created with a purpose and that this purpose was reflected in the greater number of satellites for the more distant planets. "Mercury and Venus, the planets nearest the sun, have no such attendants: the Earth has one," he wrote. And concerning Uranus: "It does not appear at all probable that he has a ring, like Saturn; but he has at least five satellites which are visible to us, ... and we believe that the astronomer will hardly deny that he may possibly have thousands of smaller ones circulating about him." An advocate of physico-theology, Whewell maintained the old-fashioned view that "the satellites are placed in the system with a view to compensate for the diminished light of the sun at greater distances." Mars was obviously an exception to the rule, but rather than suggesting that the planet had one or more undiscovered moons, he argued that "no one, ... will, by one anomaly, be driven from the persuasion that the end which the arrangements of the satellites seem suited to answer is really one of the ends of their creation."[35]

At the time Whewell wrote his *Astronomy and General Physics*, he did not harbour doubts concerning the existence of extraterrestrial beings. That only came later, with his *Of the Plurality of Worlds* published in 1853, a controversial work in which he launched a sharp attack on ideas of pluralism.

A member of the Royal Astronomical Society and since 1872 editor of its journal, the *Monthly Notices*, the British-American amateur astronomer Richard Anthony Proctor was a prolific author of both popular and technical works.[36] His greatest success, *Other Worlds than Ours*, published in London 1870, was written against the background of the controversy between Brewster and Whewell. Proctor favoured Brewster's pluralism, if in a more moderate version: He believed Venus and Mars to be inhabited, but neither the Sun nor the Moon; he also doubted if there were presently advanced life on the giant planets. In later works, published between 1874 and his death in 1888, he moved closer to the position of anti-pluralism advocated by Whewell.

Like so many earlier astronomers, Proctor was impressed by the many analogies between Venus and the Earth. "Had Venus but a moon as the earth has," he wrote, "we might doubt whether, in the whole universe, two orbs exist which

---

[35]Quotations from Whewell 1856, pp. 113–114. It was generally believed at the time that William Herschel had discovered six moons around Uranus. See further in Chapter 7.

[36]On Proctor (1837–1888), see Ranyard 1889. Apart from his astronomical works, Proctor also has a place in the history of cosmology. In *Other Worlds than Ours* he conceived of a hierarchic model universe of such a kind that, even if the universe were filled with an infinitude of stars, the total amount of starlight received on Earth would be quite small. In this way he escaped Olbers' paradox without introducing interstellar absorption. This kind of cosmological model was later developed by the Swedish astronomer Carl Charlier (1862–1934) and is sometimes known as Charlier's hierarchic universe.

are so strikingly similar to each other."[37] He then went on to consider what he called "one of the most perplexing enigmas that has ever been presented to astronomers," namely the question of a satellite encircling Venus. Proctor admitted that contemporary astronomers had found no signs of a moon, but he nonetheless found the historical evidence in favour of its existence to be impressive.

> Are we indeed certain that Venus has no moon? The question seems a strange one, when it is remembered that year after year Venus has been examined by the most eminent modern observers, armed with telescopes of the most exquisite defining power, without any trace of a companion orb being noticed. Nor, indeed, can any reasonable doubts be entertained respecting the moonless conditions of Venus, by those who appreciate the character of modern telescopic observations; and yet, if I had begun this paragraph by stating the evidence in favor of the existence of a satellite, I believe that nearly every reader would have come to the conclusion that almost certainly the Planet of Love has an attendant orb.[38]

Referring to the observations from Cassini to Horrebow, Proctor was willing to believe in a minute satellite, such as Smyth had argued in 1844. Contrary to most earlier astronomers, Proctor did not see the relevance of a satellite to be its light at night, but rather its function as a regulator of the tides. If Venus were supplied with a moon of the same kind as the one of the Earth, its oceans would have constant tides because of the cancellation of the gravitational pulls from the moon and the Sun. From this he concluded that "Venus has no need of lunar tides."[39] Of course, this teleological argument, reflecting his pluralist belief, did not preclude the existence of a very small satellite.

It is well known that the influential and popular philosopher Herbert Spencer was a great champion of Laplace's nebular hypothesis of the formation of the solar system. It is less well known that he applied the hypothesis to suggest a theory of planetary formation and the distribution of satellites. He developed this theory in an article of 1858 where he dismissed the traditional notion, based on the argument of design, that the number of satellites increased with the distance of the planet. "What is to be said of Mars, which, placed half as far again from the Sun as we are, has yet no moon?"[40] It was, of course, possible to reverse the argument and suppose that Mars did have a moon, but this possibility he chose to disconsider.[41]

[37]Proctor 1896, p. 84.

[38]Ibid., p. 85.

[39]Ibid., p. 86.

[40]Spencer 1858, p. 86. First published in *Westminster Review* **70** (1858), 185–225 and included in the 1891 edition of his *Essays* (Spencer 1891). On Spencer's theory, see Brush 1996, pp. 50–53.

[41]Lardner 1860, p. 213, found it "not altogether improbable that a satellite of Mars may yet

According to Spencer, the ratio of gravitational to centrifugal force determined the tendency to form satellites: Those planets which rotated with the greatest speed should have the most satellites. He stated the ratio of the various planets, and their number of known satellites, in the form of a table:

Mercury	Venus	Earth	Mars	Jupiter	Saturn	Uranus
362	282	289	326	14	6.2	9
0	0	1	0	4	8	4 or 6

The smaller the figure, the more satellites. In this way Spencer explained that Mars was a little less likely than Earth to have a satellite, but he also noted the "seeming anomaly" that his theory predicted that Venus should have one (since its ratio between gravity and centrifugal force was slightly less).

The anomaly might be apparent only, as it relied on what possibly was an inaccurate value of the diameter of Venus. So as to safeguard his theory, Spencer left the door open to the possibility that Venus did have a moon: "Not a few astronomers have asserted that Venus *has* a satellite. Cassini, Short, Montaigne of Limoges, Roedkier, and Montbarron, professed to have seen it; and Lambert calculated its elements."[42] Although he did not really believe in the satellite of Venus, nor did he think it was ultimately disproved. Spencer's argument on the number of satellites, including the possibility of a Venus moon, was adopted by the American spiritualist Hudson Tuttle, who incorporated it in his *Arcana of Nature*, a work published in 1860.[43]

What was perhaps the only effort of the mid-nineteenth century to observe the phantom moon of Venus was made by John Craig, an eccentric English amateur astronomer and retired country clergyman. Craig not only dreamed to construct the world's largest achromatic telescope, he actually did it. The enormous instrument had a total length of 85 feet and was placed on Wandsworth Common in London. (The tube of Lord Rosse's famous mirror telescope of 1844 – the "Leviathan of Parsonstown" – had a length of 58 feet.) Completed in September 1852, Craig's intention was to use it for studies of Saturn's ring system and to decide the old question of Venus' moon. Understandably, the monster refractor attracted a great deal of local attention. Lord Rosse visited the observatory and politely expressed his admiration of it. The telescope was described in the *Illustrated London News*, where it was praised for its powers as a measuring instrument. "Now that the instrument is adjusted, Mr. Craig wishes the Planet Venus to be examined, for he hopes to settle the question as to whether she has a satellite or

be discovered." The argument was the traditional one of a correlation of the number of satellites with the distances of the planets from the Sun.

[42] Spencer 1858, p. 87.

[43] H. Tuttle, *Arcana of Nature*, 2 vols. (Boston: Berry, Colby & Co., 1860–63), reprinted in Tuttle and Denmore 2003, where the reference to the satellite of Venus is on p. 169.

not, and we need not say what an advantage the solution of this fact would be to science."[44]

Craig's expensive brain child was a singular failure. Only a few years after its completion it was dismantled and removed, apparently without having contributed anything of value to the science of astronomy. It is unknown if it helped Reverend Craig to form an opinion about the satellite of Venus.

5.3 The Venus moon reconsidered

The most detailed reconsiderations of the Venus moon prior to the 1880s were undertaken by two obscure German amateur astronomers, C. Haase and F. Schorr.[45] Haase's involvement in the matter had its origin in the discovery of the anomalous motion of Mercury's perihelion and the early attempts to explain it in terms of the perturbations caused by an intramercurial planet. (As mentioned in section 2.1, the existence of intramercurial planets had been suggested by Otto von Guericke as early as 1672.) The leading astronomer in the discovery of the anomaly, France's celebrated Urbain Jean Joseph Leverrier, suggested in 1859 that it could be accounted for if there were an additional mass – either a planet or a series of smaller bodies – between Mercury and the Sun. Soon thereafter Edmond Lescarbault, a physician and amateur astronomer, announced that he had observed what Leverrier took to be the new planet, soon known as Vulcan.[46]

The danger of mistaking sunspots for unidentified celestial objects, whether a satellite of Venus or an intramercurial planet (or a second moon of the Earth), was well known in the nineteenth century. In connection with the hypothesis of Vulcan, numerous observations of suspicious spots were reported. Were they new planets

[44] *The Illustrated London News*, 28 August 1852, which includes an illustration of the telescope and its associated brick tower, 61 feet in height. For details about Craig (1805–1877) and his telescope, see Steel 1982, Smye-Rumsby 2004 and the website
`http://homepage.ntlworld.com/greg.smyerumsby/craig/index.html`.

[45] Neither Haase nor Schorr is listed in Poggendorff's *Biographisch-Literarisches Handwörterbuch* and they are also absent from Hockey 2007 and from Royal Society's *Catalogue of Scientific Papers* (1800–1900). A Hannoverian civil servant, Haase published in 1857 and 1861 a couple of observations in the *Astronomische Nachrichten.*

[46] The literature on Vulcan and the Mercury anomaly is extensive. See Baum and Sheehan 1997 and the sources mentioned therein. On the anomaly as a problem for Newton's law of gravitation, and its role in the formation of Einstein's theory of general relativity, see Roseveare 1982. There are certain similarities between the observations of Vulcan and the Venus moon apart from both being spurious. As Baum and Sheehan observes, "Like the phantom satellite of Venus of the 17th and 18th centuries, Vulcan managed to reappear just often enough to maintain a shadowy existence among true believers" (ibid., p. 168). A main difference between the two cases is that whereas fundamental physics (Newton's law of gravitation) was involved in the case of Vulcan, the satellite of Venus had no connection to theory.

passing in front of the Sun, or were they merely sunspots? The Zurich astronomer Johann Rudolf Wolf quoted in his *Mittheilungen über die Sonnenflecken* of 1859 several observations of possible planetary bodies, including Scheuten's observation of 1761 of the supposed satellite of Venus. To his mind, planet Vulcan belonged to the same category as the discarded moon of Venus. As the British astronomer Richard Carrington was quick to point out, whatever the worth of Scheuten's observation, the object could not have been an interior planet.[47]

The result of the apparent discovery of an intramercurial planet gave rise to a Vulcan-mania in astronomical circles and a general interest in new celestial bodies in the interior part of the solar system. Infected by the Vulcan-mania, Haase thought that perhaps some of the observations of the Venus satellite might have been due to one or more planets (these being asteroids or not) masquerading as a moon. He suggested that this was probably the case with some of Montaigne's observations.[48]

As Haase recognized, this idea, or something close to it, was not quite new. In a report on various observations made in Mannheim in 1811, Ferdinand Adolf Freiherr von Ende, an amateur astronomer from Celle, added a postscript on Venus' satellite, "which has been disputed with such convincing arguments that, I suppose, no one believes in it." On the other hand, he found it implausible that the observations could have been caused by optical delusions and therefore suggested: "It might be possible that just at those dates one of the five new planets, Uranus etc., was located in the vicinity of Venus and had been mistaken for its satellite. A rough estimate would suffice to substantiate or disprove this idea, made in passing."[49] Whereas Haase did not suggest the planet Vulcan as an explanation of the Venus observations, this is precisely what Mr. Arthur Blacklock from Manchester did. He knew about the Venus moon from the English translation of Humboldt's *Kosmos*, but contrary to Humboldt he was unable to believe that the observation reports – and that of the fellow-Briton James Short in particular – could be without foundation in fact. As to what fact it was founded in, he suggested that, in the case of Scheuten, "we have here a record of a transit of Vulcan."[50] As far as we have been able to ascertain, no one else felt tempted to

[47] Carrington 1860. On the attempts to find observations of Vulcan before 1859, see Baum and Sheehan 1997, pp. 157–160. Wolf 1891, vol. 1, p. 537 puts Venus' moon and Vulcan in the same basket.

[48] Haase 1863–69.

[49] Ende 1811, p. 394, letter dated 4 September 1811. Apart from Uranus, the "new planets" were the asteroids discovered in the early years of the nineteenth century: Ceres (1801), Pallas (1802), Juno (1804) and Vesta (1807).

[50] Blacklock 1868, p. 197. See also the comment by Thomas Webb in the same issue of the *Astronomical Register* where Webb admitted that the consistency of the observations in 1764 was difficult to explain (Webb 1868).

take up the suggestion.

The methodological similarity between the case of Vulcan and the one of Venus' moon was noted in *Littell's Living Age*, an American weekly magazine published between 1844 and 1941. In an article of 1876 on the planet Vulcan, the anonymous author claimed that "French writers on astronomy" still regarded the question of a Venus moon as undecided. As he pointed out, there were similarities as well as dissimilarities between the two cases of spurious objects:

> If Venus has a satellite, the smaller body cannot usually be concealed behind the planet, or (lying between the planet and us) be lost to view upon her disc. Therefore, the satellite should have been seen thousands of times by the hundreds of observers who have studied Venus, whereas there have been but twenty or thirty observations of the supposed satellite. But if there really is a planet traveling nearer to the sun than Mercury, we should only expect to see this planet on very rare occasions.[51]

Haase's lengthy account of 1863–69 was primarily a historico-critical examination of the observational evidence for and against a Venus moon, including reproductions of several of the sources. The idea that in some of the observations a distant planet may have been taken for the moon seems first to have been proposed in 1781, in a letter in which the Berlin astronomer Jean Bernoulli suggested that the recently discovered Uranus – for which he proposed the name either "Hypercronius" or "Trans-Saturnius" – might have been the culprit. In the letter, addressed to the Hungarian court astronomer Franz Xaver von Zach, Bernoulli wrote:

> Wouldn't it be possible that this wandering star could have given the occasion of the Venus moon which, as Wargentin says somewhere, has at the same time been seen to be in a hurry and in flight? It seems almost unbelievable to me that Cassini, Short, Montaigne, Baudouin and several others were unable to distinguish an object that was represented by a reflection in the eye – nothing but a shadow picture of a star – from a real star. But if one imagines that the new planet, whose visible diameter only makes up a few seconds, maybe 4–6, sides up with Venus, although admittedly this does not occur very frequently: if so, the former can quite easily be taken for a satellite of the latter, and one recognizes at once why it [the planet] can only accidentally be taken for it [the satellite], because it can appear suddenly and because of the quick motion of Venus can disappear equally suddenly.[52]

---

[51] Anon. 1876b.

[52] Letter of 10 November 1781, reproduced in part by von Zach in his *Monatliche Correspondenz*

Some twenty years later Julius August Koch, a medical doctor and amateur astronomer from Danzig, looked more closely at the possibility mentioned by Bernoulli. In 1802 he argued that what Roedkiær had observed in Copenhagen on 4 March 1764 might possibly have been Uranus, at the time still undiscovered. As to the other observations, Koch suggested that they were probably due to an as yet undiscovered planet. "I cannot deny that I strongly believe in such a planet, especially since the discovery of Pallas," he wrote. He thought there existed "a primary planet whose average distance from the Sun is almost the same as that of Venus, but which is smaller."[53] Moreover, he assumed that the hypothetical planet moved in an orbit strongly inclined relative to the Venus orbit and that part of its surface was dark. If so, the rotating planet would reflect light irregularly and "in this way all the other appearances of the satellite of Venus may be explained in a quite satisfactory way."[54]

It is unclear if Haase shared Koch's view or believed that there was convincing observational evidence for a Venus moon. As to Roedkiær, if he had unwittingly observed Uranus in 1764, he was not the first to do so. The planet had been recorded on a number of earlier occasions, only taken to be an ordinary star, by Flamsteed, Bradley and Tobias Mayer, among others.[55] Although these astronomers actually saw the planet Uranus, of course they did not discover it.

If Haase did not spell out his opinion about the satellite of Venus, Schorr was more explicit in a monograph he published on the subject in 1875, *Der Venusmond*. His interest in the subject was probably indebted to the transit of 1874, for as an introduction to this event he had written a small tract on the history of the solar parallax problem and the transits of Venus.[56] Schorr was convinced that "the earlier observations of the moon were not illusions" and that he had accumulated convincing evidence in favour of the real existence of Venus' satellite. "The Venus moon," he stated, "belongs to the citizens of our solar system; new observations,

(Bernoulli 1802). Baron von Zach (1754–1832) was court astronomer to Duke Ernst of Saxe-Gotha, founder and editor of the *Monatliche Correspondenz* and a leading figure in the search for missing planets between Mars and Jupiter (asteroids) in the early years of the nineteenth century. J. Bernoulli (1744–1807), sometimes known as Jean Bernoulli III, was the last of the famous Swiss dynasty of mathematicians.

[53]Koch 1802, p. 235, letter dated 24 August 1802. See also the review in *Allgemeine Literatur-Zeitung*, no. 54, 22 February 1803, p. 430. J. A. Koch (1752–1817), a member of the Danzig naturforschender Gesellschaft, wrote several papers on variable stars, asteroids and other subjects. In 1818 he published a table of variable stars. The asteroid Pallas was discovered by Olbers on 28 March 1802 and Gauss found its orbital inclination to be 34°.

[54]Ibid., p. 236.

[55]For a count of 23 pre-discovery observations of Uranus, from 1690 to 1771 (but not including Roedkiær's in 1764), see Grosser 1979, p. 41. See also Forbes 1982 and the detailed list in Alexander 1965, pp. 80–91.

[56]Schorr 1873. In this work he did not mention the possibility of a Venus moon.

DER

VENUSMOND.

UND DIE

UNTERSUCHUNGEN

ÜBER DIE

FRÜHEREN BEOBACHTUNGEN DIESES MONDES.

VON

DR. F. SCHORR,

Mitglied der naturforschenden Gesellschaft zu Danzig.

BRAUNSCHWEIG,

DRUCK UND VERLAG VON FRIEDRICH VIEWEG UND SOHN.

1875.

Figure 5.2: Schorr's little book on the *Venusmond* summarized and discussed the historical data concerning the elusive satellite of Venus. Although Schorr did not succeed in changing the status of the moon from myth to reality, his work attracted some attention and contributed to a revival of interest in the hypothetical moon.

more precise than the earlier ones, will eventually prove its existence without doubt and provide means to determine its orbit with such accuracy that is required by the present state of science."[57] The larger part of *Der Venusmond* was a detailed and historically valuable account of the development of the telescope and its use in Venus observations in particular. Schorr was an amateur astronomer, but by no means a crank. He evidently had a sound knowledge of both astronomical practice and theory, and his knowledge of the literature on Venus and the history of the

[57]Schorr 1875, p. xi.

telescope was impressive. Although of no scientific worth, his book is an important if overlooked contribution to the history of planetary astronomy.

Unable to provide any new data – because no such data existed – Schorr argued inductively that the entire corpus of observational evidence could best be explained if the Venus moon was admitted as a reality. But what about the standard objection that the moon had not been seen for more than a century? His answer, which apparently satisfied him, was that the moon reflected only very little light and probably in an irregular manner because of the particular conditions of its surface; or perhaps the bright light of Venus made its moon invisible for long periods of time. "If there are dark fixed stars, then there may also be dark planetary satellites of which the Venus moon may be one."[58] Although the existence of invisible or dark stars was generally recognized in the second half of the nineteenth century, this was hardly a satisfying answer to the puzzle that the satellite had escaped detection since the mid 1760s. Most readers of Schorr's book have probably wondered how he could advocate the Venus moon with arguments of no greater force.

As far as we can tell, *Der Venusmond* did not make much of an impact, neither in astronomical circles nor elsewhere (and that includes the later history of astronomy). However, neither was the book ignored. It was subjected to a detailed review in *Nature* by the British clergyman and astronomer Thomas William Webb who used the occasion to revive interest in the case of Venus' satellite.[59] Webb agreed with Schorr in his critique of Hell's explanation in terms of optical deceptions, but of course he did not accept the reality of the moon. As he pedagogically pointed out, "It is one thing to invalidate an opponent's conclusion – another, to establish one's own." In other words, he agreed that the attempts to explain the satellite observations were inadequate, but not that this justified the conclusion of a real satellite of Venus. Like most other commentators, Webb found it an irresistible argument that for more than a century no astronomer, from Herschel to Pietro Angelo Secchi, had seen any sign of the moon, and that although they were equipped with far better telescopes than in the past. Something had been observed in the seventeenth and eighteenth centuries, but what? Unable to come up with a satisfactory answer, Webb suggested ("though with little confidence") that some of the observations might have been due to what he called atmospheric

[58]Ibid., p. 185. On the basis of the corpuscular theory of light, the English natural philosopher John Michell (1724–1793) had put forward the idea of dark stars in 1784, and in 1844 Friedrich Wilhelm Bessel (1784–1846) inferred that Sirius was accompanied by a small unseen star. See, e.g., Clerke 1903, pp. 399–403. Eisenstaedt 1991 is a detailed analysis of the idea of dark stars from Newton to Laplace. For the connection to black holes, see also Israel 1987.

[59]Webb 1876. Another lengthy and generally positive review appeared in Bertrand 1875 (see also below). T. W. Webb is best known for his publication in 1859 of *Celestial Objects for Common Telescopes*, a classic guide for amateur astronomy for more than half a century.

illusions. With this phrase he referred to optical illusions caused by refractions and diffractions in the Earth's atmosphere, but he did not explain very clearly what kinds of atmospheric illusions he had in mind.[60]

Webb further reported that as a very young man he had observed on 22 May 1823 a shining object "exactly resembling Mercury, or a miniature Venus" at a short distance from the planet Venus. He thought that the object could not be a star but nonetheless found it "too hazardous ... to include this with the other observations of the pseudo-satellite."[61]

[60]The suggestion of atmospheric illusions or images was much elaborated in Thirion 1885 (see section 6.1).

[61]Webb 1876, p. 195.

Chapter 6

Closure: the discussion of the 1880s

Among philosophers, Venus' lack of a moon came to attract attention when the German mathematician-philosopher Gottlob Frege in 1884 published *Die Grundlagen der Arithmetik*. The foundation of arithmetics has nothing to do with either planets or moons, but Frege happened to use the moonless Venus to express the meaning of the number zero: "If I say 'Venus has 0 moons,' there simply does not exist any moon or agglomeration of moons for anything to be asserted of; ... a property is assigned to the concept 'moon of Venus,' namely that of including nothing under it."[1] Frege wanted to emphasize that zero is a property not of any object, but of a concept. Of course, in the present context this is merely a curiosity. There is no reason to assume that Frege had any interest in Venus' moon as a possible astronomical body. On the other hand, it is permissible to speculate that his example reflected the contemporary discussion of the satellite of Venus.

The 1880s saw a last revival of interest in the Venus moon, after which it largely fell into oblivion. The reasons for the revival are not quite clear, but Schorr's book and the reviews of it by Webb and Joseph Bertrand may have played a role.[2] It is also possible that the focus on Venus in connection with the transit of 1882 (and the earlier one of 1874) was among the causes. Again, in August 1877 the American astronomer Asaph Hall was working at the U.S. Naval Observatory in Washington D.C. with the largest refracting telescope in the world. On the 12th and the 17th he discovered two small objects near Mars which he interpreted as

[1] The quotation, appearing in §46 of *Die Grundlagen der Arithmetik*, has continued to attract philosophical interest.

[2] For Bertrand's review, see below. Schjellerup 1882 was motivated by Schorr's book to communicate the observation data from the Copenhagen Observatory.

moons, soon known as Phobos and Deimos. (Hall proposed the names in 1878, referring to a verse in the *Iliad.*) He immediately used their orbits to calculate an accurate mass for Mars, much in the same way as Baudouin and Lambert had sought to do with Venus a century earlier. His value – that the Sun is 3.054 million times as heavy as Mars – agreed nicely with the one found by Leverrier and his "computers" after several years of laborious calculations based on perturbation theory. (It also agrees nicely with the modern value, which is 3.100.) The discovery of the satellites of Mars was fairly uncontroversial and won quick recognition in the astronomical community. Yet it was also unexpected, for since the seventeenth century it had been commonly accepted that Mars is a lone planet.[3] On the other hand, and as we have had occasion to mention, since the days of Kepler there had been several speculations, including a few observation claims, that the planet was surrounded by one or two moons.

One might imagine that the discovery of the two moons circling around Mars caused astronomers to reconsider the question of Venus' satellite. If it turned out that planet Mars had a moon, against all expectation, why not planet Venus? However, we are not aware of any documented connection. Hall's and other astronomers' early communications on the Mars moons seem not to have made associations to the case of Venus, nor did the literature on the satellite of Venus in the 1880s refer to the satellites of Mars. The only exception seems to have been a brief remark made by Charles Young in 1893 (see section 6.3).

During the second half of the nineteenth century another controversial observation claim turned up in astronomy, namely the claim that the Earth has more than one natural satellite. This little known episode differed in many respects from the case of Venus' satellite, especially by being effectively ignored by the astronomical community (for which reason it is so little known). On the other hand, the two claims also had some features in common. For this reason, the brief and parenthetic story of the Earth's second moon is relevant to an understanding of the saga of the satellite of Venus.

---

[3]William Herschel, the German-Danish astronomer Heinrich Louis d'Arrest (1822–1875) and a few other astronomers had searched in vain for a moon around Mars. According to d'Arrest, there was little hope of seeing a Mars satellite, should it exist (d'Arrest 1865). On the discovery, see Gingerich 1970 and Dick 1988. Earlier in the century four more moons were discovered: the moon around Neptune (William Lassell, 1846), one more Saturn moon (George P. Bond, 1848) and two more Uranus moons (William Lassell, 1851). On Hall's discovery and its reception, see also Hall 1878 and *Nature* **16** (1877), pp. 397–398, 427–428.

6.1 The planet Neith

After 1875 no one seriously suggested that the satellite of Venus really existed. This implied a problem-shift, as the consensus meant that the problem now was to account for the observations in some other way. In so far as the problem was considered at all, this had been the attitude of most astronomers since the 1780s. It could of course be argued that earlier astronomers, from Fontana to Horrebow, had been fooled by their instruments and reported observations that just did not exist. But this possibility was denied by all nineteenth-century commentators who stressed that the involved astronomers were of the first rank (such as Cassini, Short and Horrebow) or at least competent and experienced observers (such as Montaigne, Roedkiær and Montbarron). Fontana's early observations were suspicious and might be explained as the result of inadequate experience with the telescopes of his day, but not so with the later observations.

To dismiss the observation claims as plain mistakes would be to throw doubt on the ability of astronomers from Cassini to Horrebow, indeed on the very profession of observational astronomy. (As early as 1728, Bianchini defended in this way Fontana's honour as an astronomer.) For example, the eminent mathematical physicist Joseph Bertrand, professor at the Collège de France and perpetual secretary of the Académie des Science, went out of his way to deny the possibility. Thus, in a review article of 1882 he emphasized that "The astronomers we have quoted, without being of the first rank, are worthy to be trusted."[4] Bertrand suggested that while some of the problematic observations might be due to false images, others were caused by asteroids in the vicinity of Venus. He had first advocated this idea in 1875, in a review of Schorr's *Der Venusmond*, in which he argued that the observations were real and most likely due to asteroids, although he left the door open for the existence of a real Venus moon. He did not believe in it, though.[5]

Camille Flammarion, the famous French astronomer and popularizer of astronomy, made his name at an early age with the publication in 1862 of *La pluralité des mondes habités*. In this and other of his many books he brought home the message that the Earth is not a privileged object in the solar system and therefore not the only inhabited "world." In a later work, the *Astronomie populaire* of 1880, he eloquently spoke of the consequences of the Earth and Venus being so similar:

> Of what nature are the inhabitants of Venus? Do they resemble us in physical shape? ... All that we can say is that organized life on Venus must be little different from terrestrial life, and that this world is one of

[4]Bertrand 1882, p. 203. Similarly in Bertrand 1875, p. 458.
[5]Bertrand 1875.

Figure 6.1: An artist's impression of Venus and its satellite, as included in Flammarion's journal *L'Astronomie* (Bertrand 1882, p. 201).

> those which resembles our own most. ... The only scientific conclusion we can draw from astronomical observation is that this world differs little from ours in volume, in weight, in density, and in the duration of its days and nights. ... It should, then, be inhabited by vegetable, animal and human races but little different from those which people our planet.[6]

But in spite of all similarity between the two planets, Flammarion did not believe in Venus having a moon of its own. He mentioned the asteroid hypothesis as a possible explanation of the observations of the satellite, which strangely had not been seen since the 1760s:

> Since then, nobody has seen it. Has it felled down on the planet? That is the last possible hypothesis. Have all these observers seen badly?

[6]Flammarion 1880, pp. 463–464.

> Certainly not. How then explain these appearances and this disappearance? Probably, Venus is found in these epochs to move in front of one of the many small planets found between Mars and Jupiter.[7]

Perhaps significantly, Flammarion relegated his reference to the satellite of Venus to a footnote and he did not follow up his suggestion, which had more the character of a side remark than a scientific hypothesis. Inspired by Bertrand's article, he followed up on the subject in *Les terres du ciel* of 1884, now suggesting that the observations could be explained by a combination of the asteroid hypothesis and the optical illusion hypothesis of Hell. As to the real existence of the moon, he dismissed it as highly improbable.[8]

Figure 6.2: The French astronomer and writer of popular astronomical works, Nicolas Camille Flammarion (1842–1925), was obsessed with the idea of extraterrestrial life. Although he took the existence of Venusians for granted, he did not believe that Venus had a small moon.

[7]Ibid., p. 464. As mentioned above, the asteroid hypothesis had previously been suggested by von Ende and Haase. On Flammarion as a pluralist, see Crowe 1999, pp. 378–386, 410–433.

[8]Flammarion 1884, pp. 262–266.

In spite of Thomas Webb's objections against Hell's hypothesis of instrument-produced ghost images, this kind of explanation enjoyed wide recognition. It was easy to comprehend and sufficiently broad to encompass almost all the erroneous observation claims. As Bertrand pointed out, it was supported by recent observations by the British astronomer William Frederick Denning, who in 1881 had observed *two* crescents when he looked upon Venus. "The reputed observations of a satellite of Venus, described in astronomical text-books, immediately recurred to me," Denning wrote. When he rotated the eyepiece, the relative positions of the two objects remained the same. Denning readily came up with an explanation in terms of reflections in the telescope. "I have no doubt," he wrote, "that the alleged observations of a satellite of Venus made in the last century were capable of a similar solution."[9]

Haase's idea to connect the observations of the Venus moon with a hypothetical planet moving within the orbit of Mercury was independently proposed by the Belgian astronomer Jean-Charles Houzeau de Lehaie in a public lecture in 1880.[10] Houzeau, who had started his career as a journalist and was equally at home with social and scientific questions, was at the time director of the Royal Observatory in Brussels, an institution founded in 1823 with the famous statistician and sociologist Adolphe Quetelet as its first director. (Although best known as a social statistician, he also did important work in astronomy and meteorology.) The successor of Quetelet, Houzeau was a scientist of considerable influence. He first addressed the question of Venus' satellite in an address of 1878 before the Royal Belgian Academy of Sciences and Letters, dealing with "Certain enigmatic phenomena in astronomy." After having accounted for the historical observations, he pointed out a number of reasons why the hypothetical satellite had rightly fallen in disfavour. Among those reasons were not only the non-observations related to the transits of 1761, 1769 and 1874, but also that the moon had not been observed for a period of 114 years. "If it is extremely faint, how to explain that the old telescopes sometimes display it, and that the powerful instruments like those used to discover the satellites of Uranus, Neptune and Mars have not revealed it to us?"[11] Apart from the possibility of optical illusions, he also mentioned that what had been taken for a moon might have been an intramercurial planet.

[9]Denning 1882, p. 111. Denning (1848–1931) was a prominent amateur astronomer who discovered four comets. Obituary notice in *The Observatory* **54** (1931), 276–283.

[10]Houzeau, who served as director of the Royal Observatory in Brussels 1876–83 and for a period was president of the Royal Belgian Academy of Sciences, was active in the Venus transit observations in 1882. His interest in the Venus moon is illustrated by his massive astronomical bibliography, which included the subject of the moon of Venus as a separate entry (Houzeau and Lancaster 1964, vol. 2, columns 1136–1137). His eventful life, which included political activity and extended stays abroad, is described in Verhas 2002.

[11]Houzeau 1878, pp. 957–958.

The Venus moon was not the only astronomical enigma that Houzeau dealt with in his address of 1878. As earlier astronomers had likened the phenomenon of the Venus moon to variable stars such as Mira Ceti, so Houzeau drew a parallel to the mysterious comet Biela, named after the Austrian Wilhelm von Biela who in 1826 succeeded in following the comet for 72 days, just long enough that its period of 6.6 years could be determined.[12] First observed by Montaigne in 1772, the comet was seen in 1846 as broken into two pieces and after its appearance in 1852 it was never seen again. But on 27 November 1872, when the Earth passed close to the orbit of the comet, there was a magnificent display of meteor showers, the remnants of the comet. Houzeau's point with the comparison may have been that the Venus moon might have been an object that once existed but had disappeared, possibly disintegrated, at some time after the 1760s. However, if that is what he meant, he did not spell out the hypothesis very clearly.

Whereas Houzeau had suggested in 1878 and 1880 that the Venus moon might have been an intramercurial planet, a few years later he realized that the hypothesis was untenable. By singling out seven historical observations, which he believed were the best attested, he found that in all these cases Venus was further from the Sun than an intramercurial planet could possibly have been, and this ruled out the hypothesis.

Although thus falsified, Houzeau thought that a modified version of the hypothesis might fare better.[13] The observations of the seventeenth and eighteenth centuries apparently occurred randomly, but by examining the precise dates, expressed in decimals of a year, he was led to suggest that they occurred in multiples of a unit of 2.96 years. For instance, Cassini's first observation took place 26.20 years after Fontana's and 14.58 years before his second observation of 1686; and Short's observation of 1740 preceded Montaigne's of 7 May 1761 by 20.50 years.

There may not seem to be much regularity in the numbers given in column 3 of table 6.1, but for the believer there is: Houzeau thought that it was no coincidence that $26.20 = 9 \times 2.91$, $14.58 = 5 \times 2.92$, $54.16 = 18 \times 3.02$, etc. Numerological reasoning of this dubious kind made him to suggest that the historical observations were of two bodies in or near conjunction, one relatively large and the other of much smaller dimensions. "There can be no question that it cannot be a true satellite," he observed, referring to the smaller body. "It follows from the observed facts that the path they follow brings them together at fixed intervals of time.

[12] Only with Biela's observation was it realized that the comet of 1826 was the same as that seen in 1805 by the great comet hunter Jean Louis Pons (1761–1831) in Marseille, and that this was the same as Montaigne's comet of 1772. The official name for the comet is 3D/Biela, where D stands for "defunct." For a brief history, see `http://cometography.com/pcomets/003d.html`.

[13] Houzeau 1884, based on an article in *Ciel et Terre* of the same year, with excerpts translated in Anon. 1884.

observations	**date**	**interval** (years)	**no. of periods**	**length of period** (years)
1 (Fontana)	1645.87	–	–	–
2 (Cassini)	1672.07	26.20	9	2.91
3 (Cassini)	1686.65	14.58	5	2.92
4 (Short)[14]	1740.81	54.16	18	3.02
5 (Montaigne)	1761.34	20.50	7	2.97
6, 7 (Montbarron)	1764.24	2.90	1	2.90

Table 6.1: Houzeau's data for the observations of the moon of Venus, which he thought was evidence for a new planet.

These paths are near one another throughout their entire length ... [and] these conditions can only be satisfied by imagining two orbits sensibly concentric and of radii which differ but very little."[15] One of the bodies was of course Venus, and he proposed to call the minor one Neith, named after an Egyptian goddess for hunting and warfare. He found that the new planet moved a little exterior to Venus and with a period of revolution of about 283 days or 0.78 years.

The planet named Neith was much more short-lived than the unnamed satellite of Venus. In fact, the hypothesis was shot down immediately after it had been proposed. Not only was the selection of the seven observations questionable and the numerological basis of the hypothesis objectionable from a methodological point of view; the suggestion also suffered from internal inconsistency, such as pointed out in a review in *The Observatory.* The anonymous reviewer – probably Edward Walter Maunder, the journal's editor – found the asteroid hypothesis to be implausible and also had little faith in "the vague and unsatisfactory resource of ascribing as many of the observations as possible to false images." Perhaps, he wrote, "it is better to be content with things as they are, and to leave the 'satellite' of Venus, at least for the present, as an unsolved 'astronomical enigma'."[16]

The American astronomer Charles Young, professor at Princeton University and a leading solar spectroscopist, agreed that Houzeau's hypothesis was untenable, but disagreed that the case of the Venus moon was an enigma. As far as he was concerned, the case was solved and the solution was to be found in the old analysis of father Hell. "There can be little doubt," he wrote, "that all the Venus

[14]In the case of Short's observation of 1740, Houzeau mistakenly used the old time style as used by Short. If the new one is used, as it should, the interval of time from Cassini's observation becomes 54.19 years and the length of the period 3.01 years.

[15]Houzeau 1884, p. 285.

[16]Anon. 1884, p. 226. A more elaborate rejection of planet Neith appeared in Stroobant 1887a, pp. 15–18.

satellites so far observed are simply ghosts due to reflections between the lenses of the telescope, or between the cornea of the eye and the eye lens."[17]

Julien Thirion, a Belgian Jesuit astronomer and physicist, reconsidered the problem of the Venus moon in a paper of 1885, ten years after Schorr's monograph had appeared. Having no faith in the real existence of the satellite, he found that neither the asteroid hypothesis nor the false image hypothesis à la Hell were able to account for all observations. As to Houzeau's recent suggestion of the planet Neith, Thirion pointed out that it was based on an arbitrary selection of data. If all observations were included, Houzeau's numerology just did not work. Therefore, some other explanation was needed. Inspired by earlier observations of false images of the Sun, so-called mock suns (also known as sun dogs or parhelia), Thirion suggested that the apparition of a Venus moon might belong to the same class of phenomena.[18]

The phenomenon being mentioned by ancient authorities such as Aristotle, Cicero and Pliny the Elder, mock suns had been reported long before the invention of the telescope. In the seventeenth century they were known from observations made by Jean Dominique Cassini, Christopher Scheiner and Johannes Hevelius, among others, and Descartes and Huygens had attempted to explain from optical theory the strange halo-like phenomenon.[19] A mock sun is an atmospheric optical phenomenon, a luminous cloudy patch which appears in ice clouds at either side of the Sun. Sometimes it is round, like the real Sun, but this is not always the case; and in some cases the mock sun appears as provided with a tail. A similar phenomenon of false images produced in the atmosphere exists for the Moon, in which case one speaks of paraselenae. For example, Hevelius observed in 1660 a triple picture of the Moon, with two mock moons shooting out long whitish beams.

The appearance of parhelia is caused by diffraction by ice crystals in the upper atmosphere, and it occurs when the Sun shines through a cloud composed of hexagonal ice crystals falling with their principal axes vertical. If the crystals are randomly oriented, a halo is observed. Parhelia may be white or coloured, depending on the role played by reflection and refraction. Other crystals than ice may cause similar phenomena. Thus it is known that the atmospheres of Saturn, Jupiter and Uranus contain ammonia, methane and other compounds which can

[17] Young 1886, p. 249. See also Lynn 1887a, p. 74.

[18] Thirion 1885. A frequent contributor to the Catholic journal *Revue des Questions Scientifiques*, Thirion (1852–1918) published on a variety of astronomical and physical topics, including the history and philosophy of science.

[19] Hevelius 1674. Aristotle referred to the phenomenon in his *Meteorologica*. For a modern example of a mock sun, looking like a satellite accompanying the Sun, see *Sky and Telescope* **110**, no. 7 (2005), p. 128. For a history of mock suns until the mid-eighteenth century, see Priestley 1772, pp. 613–630. There is no modern history of the subject.

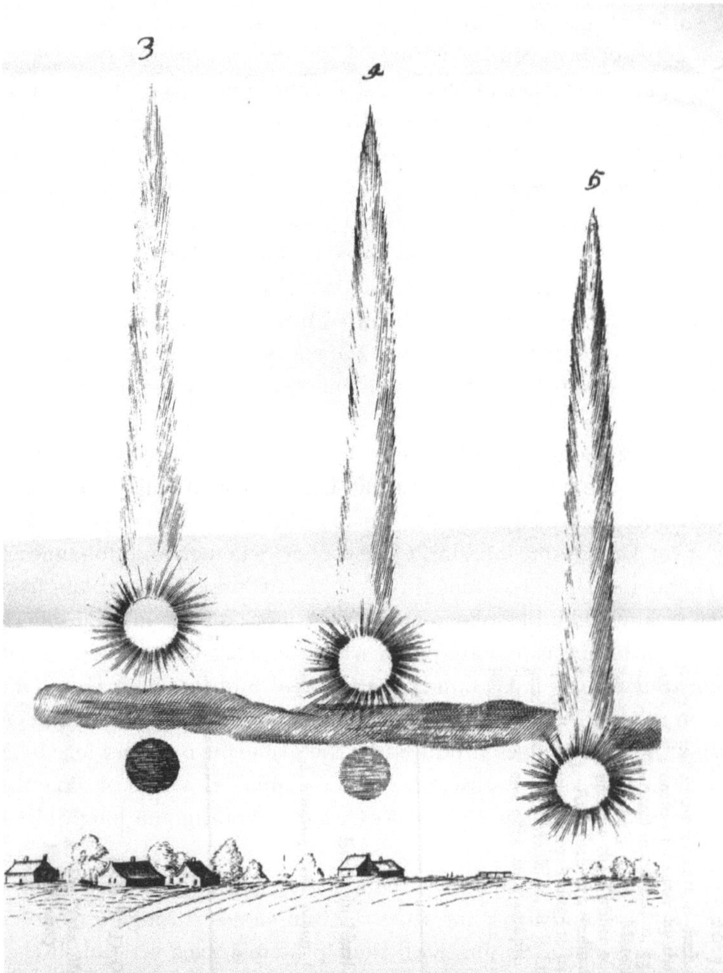

Figure 6.3: An unusual parhelion or mock sun reported to the Royal Society by Hevelius. On 5 February 1674 the Sun was seen with a cloud beneath it. The Sun lanced out long rays and beneath the cloud there appeared a mock sun of almost the same size as the real Sun. The spurious sun disappeared as the Sun descended below the cloud. From Hevelius 1674.

produce halos of the same kind as parhelia. However, this was not known in the mid-nineteenth century.

The modern explanation of halos and parhelia is due to the French physicist and crystallographer Auguste Bravais, professor at the École Polytechnique, who in a lengthy treatise of 1847 offered a theory of the phenomena which is essentially the one accepted today. Thirion adapted Bravais' theory to explain the observations of the Venus moon.[20] His explanation was thus based on optical illusion, but contrary to Hell's it referred the mechanism to the atmosphere, not to the telescope and the eyes. Contrary to earlier commentators, Thirion paid much attention to Fontana's observation of 11 November 1645, in which the Neapolitan astronomer had seen *two* moon-like objects. This could be explained on the ice crystal hypothesis, but not on any of the other hypotheses. According to Thirion, his theory was superior to the alternatives, indeed "plausible and likely to be true." However, it was ignored by most astronomers who saw no reason to reopen the case of the satellite of Venus, believed to be just a mistake.

Many of the writings on the satellite of Venus in the 1880s were due to Belgian astronomers, who seem to have taken a particular interest in the subject. This interest – as manifested in works by Houzeau, Stroobant and Thirion – may in part have been indebted to observations made at the Brussels Observatory in February 1884. On the third of this month E. Stuyvaert saw on the disk of Venus a luminous point that looked like one of Jupiter's satellites passing in front of this planet. Nine days later, his colleague Leopold Niesten observed an unidentified object close to Venus, apparently a small star surrounded by a faint nebulosity. None of the Belgian astronomers suggested that they had actually observed the satellite of Venus and apparently they did not report their observations in the scientific literature.[21] Nonetheless, they were known by their fellow astronomers in Brussels.

The revival of interest in the Venus moon in the 1880s was not restricted to Belgium. We are aware of only one astronomer, S. J. Lambert from Auckland, New Zealand, who watched the transit of 1874 with the moon in his mind. He reported that he found no trace of the satellite.[22] The Irish astronomer William Edward Wilson observed the 1882 transit of Venus from his observatory in Daramona near Cornacausk, Ireland.[23] Primarily looking for traces of absorption of the solar

[20]Thirion 1885; Bravais 1847. For criticism of Thirion's hypothesis, see Stroobant 1887a, p. 15.

[21]A brief account of the observations of Stuyvaert and Niesten is given in Thirion 1885, p. 46 and Stroobant 1887a, pp. 8–9. Stuyvaert and Niesten believed that they had seen rapid displacements, relative to Venus' terminator, of both bright and dark spots, and for this reason they supported the short rotation period of 23 hours rather than the long one proposed by Schiaparelli. See *Monthly Notices of the Royal Astronomical Society* **52** (1892), 281–282.

[22]According to *The Amateur Astronomer* **18**, no. 1 (1958), 3–6.

[23]On Wilson (1851–1908) and his astronomical work, see the obituary in the *Proceedings of the Royal Society A* **83** (1910), iii–vii, and also Warner 1977.

spectrum in the atmosphere of the planet, he also tried to spot a satellite of Venus. However, none was found.

At Crowborough, Sussex, Leeson Prince observed the transit, noting that a halo of yellowish light surrounded the planet when it was fully on the Sun's disk. Aware of the discussions of a possible moon of Venus, he briefly reported that "There was no appearance of a satellite."[24] The following year he published his observations in a booklet which included a translation of Baudouin's account of 1761 of the supposed discovery of the satellite of Venus.[25] Although Prince did not endorse the real existence of the moon, nor was he inclined to write it off for good. This was also the opinion of W. T. Lynn, who in the *Observatory* suggested that "we are rather, it seems to me, in the same position on the question as we were with regard to a satellite or satellites of Mars before the year 1877."[26] Lynn admitted that no satellite of Venus had ever been actually seen, but not that this was conclusive evidence against a very small moon revolving around the planet.

Another amateur astronomer who apparently took an interest in the possibility of a moon of Venus was Carl Venceslas Zenger, a professor from Prague who was a member of the Royal Astronomical Society and also of the Astronomical Society of the Pacific. Zenger was interested in the Venus transit and wrote a couple of papers in the *Monthly Notices* in which he discussed the use of photography in the coming transit. While observing Venus in October 1876, he noted "a very brilliant patch near the southern horn some two seconds or less from the terminator in the dark parts of the disk, and this was still visible for two days, when the terminator had passed it, shining like a brilliant star in the surrounding part of the illuminated disk."[27] The phenomenon he observed was of the same kind as the one reported eight years later by Stuyvaert in Brussels. Zenger's interest in the Venus moon is indicated by a paper he read before the Royal Astronomical Society on 8 March 1889, entitled "On the satellite of Venus and its revolution."[28] Unfortunately he did not publish the paper, whose content is unknown.

6.2 Stroobant's solution

In part inspired by Houzeau's ideas, in 1887 the case of the satellite of Venus was reinvestigated in a most thorough manner by another Belgian astronomer, the

---

[24] Prince 1882, p. 65.

[25] Prince 1883, reviewed in *The Observatory* **6** (1883), 160. As pointed out in section 3.1, this was a reproduction of John Bevis' translation originally appearing in the *Mathematical Magazine* in 1761.

[26] Lynn 1884, p. 231.

[27] Zenger 1877, p. 461. See also Baum 1999.

[28] Mentioned in *Monthly Notices of the Royal Astronomical Society* **50** (1890), 253.

young Paul Stroobant, who at the time worked as a voluntary assistant at the Royal Observatory at Uccle near Brussels.[29] Stroobant was familiar with most of the literature on the subject, including Schorr's book of 1875. Systematically discussing the various hypotheses, he not only dismissed the possibility of a real Venus moon but also the hypothesis of an intramercurial planet as well as Houzeau's suggestion of the planet Neith. As to the other hypotheses – based on false images, asteroids or atmospheric mirages – he found them to be either too vague or insufficient to account for all the observations. Stroobant's favoured alternative was that the supposed moons had been faint stars close to the line of sight of Venus.[30] His method was essentially the same which Hevelius had used 230 years earlier to refute Rheita's claim of having discovered five additional satellites orbiting Jupiter (section 2.3). To test his idea, Stroobant made use of the *Bonner Durchmusterung* catalogue of about 320,000 stars that had been completed in 1859 by Friedrich Argelander and his assistants. By means of the catalogue he identified stars which at the time of the historical observations were very close to the reported locations of the Venus moon. And, indeed, he found an almost perfect match.

To mention but one example, on 4 August 1761 Roedkiær had first noticed an object that he assumed was the Venus moon, but then he saw another nearby star-like object which he instead took to be the moon. From Roedkiær's data Stroobant found the coordinates of the two objects and compared them with the reductions from Argelander's star catalogue. He found the match to be convincing in both cases, leading to an identification of the first object with 64 Orionis and the second – the supposed moon – with 62 Orionis. Likewise, the satellite that Horrebow saw on 3 January 1768 agreed nicely with the position at the time of Theta Librae, a fifth-magnitude star.

Not all the identifications fitted equally well, and in the case of Cassini's observations, and also the one of Mayer in 1759, Stroobant was unable to point to a star in the catalogue. But in these cases he suggested that an error had been made in the dates, although his arguments for this suggestion were not entirely convincing. In a few other cases he fell back on the optical illusion hypothesis. As to Scheuten's observation of 1761, Stroobant thought that the German had seen only a sunspot and taken it to be a moon. (In fact, in Scheuten's letter to

[29]Remarkably, at the time Stroobant presented his work to the Royal Belgian Academy, he was only 19 years old. Two years later he earned a doctorate in physics and mathematics from the University of Brussels and then embarked on a distinguished career in astronomy which made him a professor of astronomy (1896) and director of the Royal Observatory (1925). Obituary notices in *The Observatory* **59** (1936), 349–352 and *Astronomische Nachrichten* **260** (1936), columns 175–176.

[30]Stroobant 1887a. In addition to this work, which included many excerpts from the historical sources, he published shorter versions, such as Stroobant 1887b, Stroobant 1887c and Stroobant 1888.

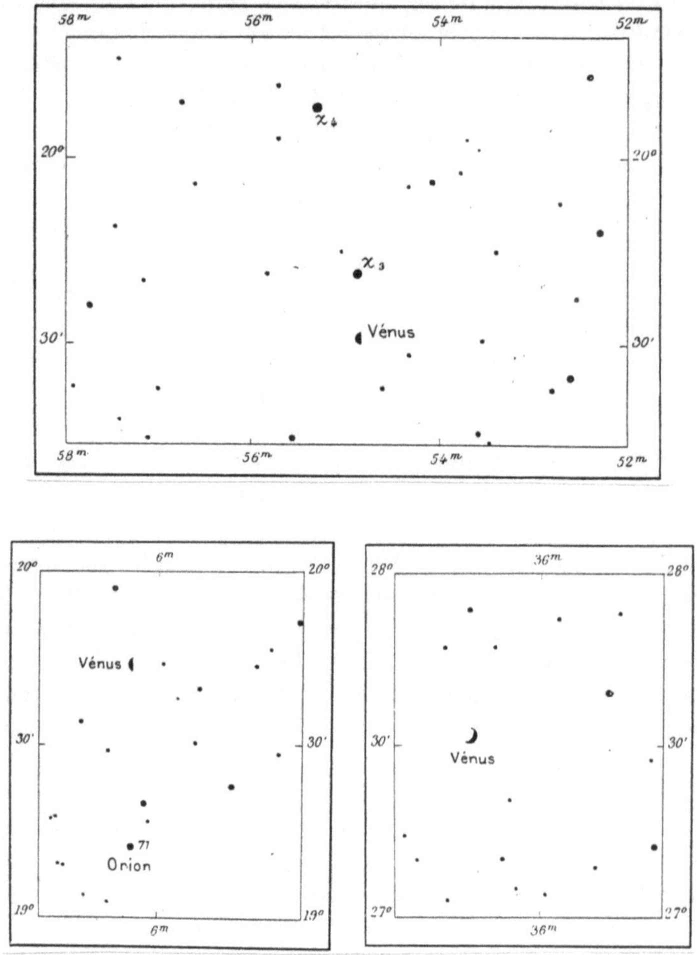

Figure 6.4: Three examples of Stroobant's star maps, from Stroobant 1887b. The image above illustrates Roedkiær's discovery claim of 4 August 1761, Venus being shown with its phase. Above Venus is the star Theta Orionis (symbol χ_3), while the symbol χ_4 denotes the location of the alleged satellite. The two smaller images refer to Montaigne's observation of 11 May 1761 (left) and Roedkiær's of 18 July 1761.

Lambert he was careful to point out that the object did not look like a sunspot at all, but this Stroobant chose to ignore.) Whatever these exceptions and difficulties, he considered them to be unimportant to his main conclusion: "In brief, we can say that the satellite of Venus does not exist, and when there was no false image or optical illusions, we find for the best observations a star corresponding almost exactly to the different observed positions."[31] Stroobant's somewhat cavalier conclusion can easily be criticized, but since it was accepted by nearly all astronomers this is beyond the point. Apart from its scientific arguments, his memoir was also valuable from a historical point of view, as it reproduced in original language the central parts of the relevant sources, from Fontana to Horrebow.

Disseminated in shorter versions in widely read journals such as *Nature, Sirius, La Nature, Astronomische Nachrichten, l'Astronomie* and *Bulletin Astronomique*, Stroobant's work soon became generally known among astronomers, most of whom had no problem with accepting his explanation as the final word on the issue. According to *Nature*, "Mr. Stroobant has fairly cleared up the mystery which has perplexed astronomers so long."[32] Hermann Klein, the editor of *Sirius*, thought likewise: "Stroobant has succeeded in solving the enigma of Venus' satellite and conclusively cancelled it from the list of astronomical problems."[33] In his *Text-Book of General Astronomy* Charles Young accepted the explanation, but cautiously added: "It is not, however, impossible that the planet may have some very minute and near attendants like those of Mars, which may yet be brought to light by means of the great telescopes of the future, or by photography."[34] A few other astronomers, while admitting the value of Stroobant's analysis, found that it did not fully explain the history of the Venus moon and that the case was only partially solved.[35] As mentioned, no one argued in favour of the moon.

More or less ostracized from the scientific literature, the satellite of Venus turned up in the science fiction literature of the late nineteenth century. Well versed in the history of astronomy, in one of his novels the famous French novelist Jules Verne referred to the mysterious satellite. In *Hector Servadac*, a science fiction novel of 1877, the hero and his team experience a marked change in the orbit of the Earth, bringing it dangerously close to Venus and the hot Sun. They

[31]Stroobant 1887b, p. 457.

[32]Anon. 1887a, in which exception was taken to one case, the observations made by Roedkiær in March 1764. A similar appreciation appeared in Anon. 1887b.

[33]Klein 1887, p. 249. Again, in a review by "R. R." in *Bulletin Astronomique* **4** (1887), 473–475: "Mr. Stroobant has succeeded to dissipate the mystery which envelopes this enigmatic satellite of Venus, and to destroy a legend that menaced to perpetuate itself" (p. 475). See also Wilson 1887. For a similar evaluation, of a somewhat later date, see Macpherson 1906, p. 88.

[34]Young 1893, p. 331. The possibility was also mentioned in Russell, Dugan and Stewart 1926, p. 320, apparently taken over from Young (it was a revision of Young's textbook *Manual of Astronomy*, first published 1902).

[35]E.g., Wolf 1891, vol. 1, p. 537.

are able to see clear evidence of high mountains and also "the seven spots, which, according to Bianchini, are a chain of seas." Fortunately no catastrophe ensues, and before the distance between the two planets begins to increase again, they have the opportunity to take a closer look at Venus: "The proximity to Venus had been close enough to demonstrate that beyond a doubt the planet has no moon or satellite such as Cassini, Short, Montaigne of Limoges, Montbarron, and some other astronomers have imagined to exist. 'Had there been such a satellite,' said Servadac, 'we might have captured it in passing.' "[36]

Capitalizing on the success of Verne, the French authors Georges le Faure and Henry de Graffigny published in 1888 a four-volume science fiction novel, the *Aventures extraordinaires d'un savant Russe*, which tells the story of a daring exploration of the solar system and the space beyond it.[37] The group of French and Russian astronauts fly to the Moon and from there, on a spacecraft propelled by solar pressure, to Venus. Then they proceed to Mars, passing on their way its two moons, and eventually they pay a visit to Jupiter. The plot and style of the novel is Vernerian throughout. On their way to Venus, one of the scientists asks about the disputed satellite. Does it really exist? Mikhail Ossipoff, the novel's Russian space explorer, responds:

> Many astronomers have thought to have seen the satellite you speak about; for my part, and in spite of the numerous tracts published on the subject, I persist to consider its existence to be problematical. On the other hand, you may reply if it is not difficult to admit that scientists such as Cassini, Horrebow, Short and Montaigne have seen wrongly or have been victims of an optical illusion. ... I think that only two explanations are possible: either that they took a small planet passing the field of sight for Venus' satellite; or that the satellite, which must be very small, is not visible from the Earth because of some quite exceptional conditions.

Another of the company, named Gontran, mentions the possibility that the satellite no longer exists but has fallen into the planet. This Ossipoff finds to be a reasonable hypothesis, for "no law of nature forbids that such a phenomenon can occur."

---

[36] *Hector Servadac, voyages et aventures à travers le monde solaire* was published in Paris in 1877, and the following year an English translation appeared as *Hector Servadac, or the Career of a Comet.* Later editions were published with different titles, such as *Off on a Comet, or Hector Servadac.* The work is available online, see `http://ebooks.adelaide.edu.au`.

[37] Faure and Graffigny 1888, which includes a preface by Flammarion. We have used the online edition `http://www.gutenberg.org/etext/2462`, where the quotation appears on p. 65.

6.3 The second moon of the Earth

The Earth has its Moon, but does it possess more than one? Or does the Moon perhaps have its own moon?[38] In a certain sense, the Earth once had more than one satellite, or at least so it was believed before the acceptance of the Copernican system. According to the traditional world view of Aristotle and Ptolemy, the Sun and the planets revolved around the Earth in the same way as the Moon, only at much greater distances. When Copernicus' heliocentric theory became generally adopted in the seventeenth century, the number of satellites shrank drastically, from seven to one. But of course, in the present context – relating as it does to the post-Copernican era – a satellite means a body revolving around any of the planets recognized as such according to the heliocentric system.

The suggestion that there is more than one moon encircling the Earth may first have been made by Thomas Clap, the first president of Yale College. In a posthumously published work of 1781, *Conjectures upon the Nature and Motion of Meteors*, Clap argued on the basis of observations of three meteors that they were "terrestrial comets" which would come as close to the Earth as about 40 km.[39] The idea of terrestrial comets orbiting elliptically around the Earth, as if they were satellites, had but few adherents, but it continued to attract some attention well into the nineteenth century. Thus, in 1811 John Farley suggested from observations of meteors that there existed "an almost infinite number of *satellitulæ*, or very small moons, constantly revolving round the Earth, in all possible directions, and appearing only during the very short time that they dip into the upper part of the atmosphere each time that they are in perigee."[40]

The French astronomer Frédéric Petit, director of the Toulouse Observatory, specialized in the study of meteors and bolides (fireballs), that is, bright meteors that appear to explode rather than just fizzle out. In two papers in the *Comptes Rendus* of the Académie des Sciences of 1846–47 he discussed observations of bolides appearing on 21 March and 23 July 1846. Calculating the tracks of the orbits, he concluded that they were small satellites of the Earth. "This apparition," he wrote, referring to the second bolide, "has brought me to the conclusion that, most likely, the Moon is not the sole satellite of the Earth, but that other bodies of much smaller volume encircle our planet."[41] He found that the supposed meteoric satellite of 21 March moved in an orbit of eccentricity 0.38, with a closest distance to the surface of the Earth being 11.5 km; the corresponding values for the bolide

---

[38] On the second moon of the Earth, see Ashbrook 1955 and Bakich 2000, pp. 145–149. See also Schlyter 2003 and the identical version in `http://www.nineplanets.org/hypo.html`. The present account relies on Kragh 2008.

[39] Greene 1954, pp. 348–349; Burke 1986, p. 23.

[40] Farley 1811, p. 286.

[41] Petit 1847, p. 261. For the earlier paper, see Petit 1846.

of 23 July were 0.45 and 6 km.

Petit's discovery claims were generally met with silence. However, Urbain Leverrier, who had recently acquired fame for his prediction of the planet Neptune, felt provoked to respond. In a long and detailed communication of 1851 to the *Comptes Rendus*, he flatly rejected the Toulouse astronomer's idea of a second moon in the shape of a bolide, which he thought was nothing but a chimera. According to Leverrier, Petit's "strongly hazardous" conclusion was stated with an unwarranted confidence, as the uncertainty in the data had not been taken properly into account and errors due to air resistance had been ignored. "The uncertainty in the observations made of the course of the bolides permits a variety of hypotheses," Leverrier summarized. "It is unwarranted to prefer the one in which this or that bolide is a satellite of the Earth; on the contrary, this is the only hypothesis that seems to be excluded by the physical circumstances of the phenomenon."[42] Petit, who was a corresponding member of the Academy, did not respond to the attack.

Although Petit's idea of a second moon was resisted by the great majority of astronomers, it was favourably mentioned by Arago, under whom Petit had studied. The famous physicist did not actually support the idea, but he did not write it off either. According to Arago's account in his *Astronomie populaire*:

> Several astronomers are of the opinion that the bolides should be considered as satellites of our planets, as they move with extraordinary speed around the Earth and can be observed repeatedly. During the later years, Petit, the director of the Toulouse Observatory, has with great perseverance sought to find the orbits of the bolides according to the mentioned hypothesis, and he has succeeded to do so with a certain precision.[43]

Also Amédée Victor Guillemin, a French astronomer, journalist and writer of popular works on science, referred positively to Petit's hypothesis. In 1866 he published a book on the Moon, *La lune*, in which he included a brief chapter entitled "Is the Moon the only satellite of the Earth?" Taking Petit as his authority, Guillemin wrote that the new satellite, if it existed, "revolves around us with a period which does not exceed 3 hours 20 minutes, and its mean distance to the centre of the globe is 14,500 kilometres. It follows that the distance from the surface of the Earth is less than 8,140 kilometres, that is, it should be about 46 times and one third less than the distance of the Moon."[44]

---

[42] Leverrier 1851, p. 566.
[43] Arago 1854–60, vol. 4, p. 281.
[44] Guillemin 1866, pp. 192–193.

If Petit's moon was either rejected or politely ignored by the professional astronomers, it attracted a great deal of attention among amateurs and the general public. A major reason for this was that it figured in one of Jules Verne's early novels, the *Autour de la lune* of 1869.[45] The French novelist and science fiction writer had one of his characters say: "This second moon is so small and its velocity is so great that the inhabitants of Earth cannot see it. It was by noticing disturbances that a French astronomer, Monsieur Petit, could determine the existence of this second moon and calculate its elements. According to his observations, this meteorite will accomplish its revolution around the earth in three hours and twenty minutes, which implies a wonderful rate of speed."[46] There is little doubt that Verne's source of information about Petit's second satellite was Guillemin's popular book and not the scientific paper in the *Comptes Rendus*. Thus, he gave the time of revolution as 3 hours 20 minutes, as stated by Guillemin, rather than the more precise value of 12,286 seconds as given by Petit in his communication to the Académie des Sciences.

Among those who claimed to have discovered a second satellite of the Earth was one Georg Waltemath, a citizen of Hamburg, who confidently predicted that it would pass in front of the Sun on 3 February and 30 July 1898 (needless to say, it did not).[47] Moreover, based on what he thought were earlier but unrecognized observations of the moon he calculated its orbital elements. He gave the mean synodic period of revolution as 177 days, its mean distance from the centre of the Earth as 1.03 million km, and its volume and mass relative to the Moon as 1/123 and 1/80, respectively. Although Waltemath mentioned several earlier observations in favour of his claim, most of them taken from obscure sources, he was apparently unacquainted with Petit's moons of 1846–47, to which he did not refer.

According to Waltemath, the secondary satellite had been observed several times in the past, first in 1701 by Cassini senior and Maraldi who however mistook it for a sunspot.[48] He also thought that the satellite of Venus observed in St. Neot in England in 1761 was in reality the second moon of the Earth, but provided no arguments in support of the claims. This seems to be the only direct connection between the two non-existing satellites. Unconcerned about his lack of recognition among astronomers, Waltemath announced later in 1898 yet another moon of the Earth, a third one, which he said was 746 km in diameter and at a mean

[45]On Verne and astronomy, see Jacques Crovisier, "L'Astronomie de Jules Verne," a paper delivered to the *Colloque international Jules Verne: Les Machine et la science*, held in Nantes in October 2005. Online as `http://www.lesia.obspm.fr/~crovisier/JV/cro05_nantes.htm`.

[46]Quoted from the 1873 edition of *Round the Moon*, available online as `http://www.ibiblio.org/pub/docs/books/sherwood/R-II-d.htm`.

[47]Waltemath 1898, a privately published pamphlet.

[48]Cassini and Maraldi 1707.

Astronomische Mittheilungen.

Herausgeber: Dr. Georg Waltemath in Hamburg.

Aufruf an die Herren Astronomen und Freunde der Astronomie.

Ein zweiter Mond der Erde.

Die säculare Beschleunigung der Mond-Bewegung in Länge ist bekanntlich um etwa 5 Sekunden grösser, als sie nach der Störungs-Theorie aus den Einwirkungen der Planeten sein sollte, und die Versuche, diese Acceleration zu erklären, sind bis jetzt nicht gelungen. Der Unterzeichnete ist der Ueberzeugung, dass ein System kleiner Monde und Meteoriten die Beschleunigung herbeiführen. Thatsächlich ist auch am 11. Juni 1855 von Dr. Ritter, einem Lehrer der Polytechnischen Schule von Hannover, mit blossem Auge ein runder, schwarzer Körper beobachtet worden, der von **rechts nach links** an der Sonne vorüberging.

Der Unterzeichnete hat es sich zur Aufgabe gemacht, die Bewegung dieses Erd-Mondes zu erforschen und bringt nachstehend die Elemente desselben zur allgemeinen Kenntniss.

Mittlere Länge am 1. Januar 1898 Mittags Berlin 214,73°.
Mittlerer synodischer Umlauf 177,00593 Tage.
Umlauf siderisch . . . . . 119,227434 Tage.
" **drakonitisch** . . 118,003952 Tage.
" **anomalistisch** . 120,5915 Tage.
Mittlere tägliche Bewegung 3°,0194393012.
Erdnähe 1898 April 8 Mittags Greenwich.
Niedersteigender Knoten 1898 Februar 3 Abends 6,72 Uhr Greenwich in 315° 13' Länge.
Neigung 2, 5 Grad.

INSTITUT FOR DE EKSAKTE
VIDENSKABERS HISTORIE
AARHUS UNIVERSITET

Figure 6.5: Georg Waltemath's *Aufruf* to the astronomical community concerning the Earth's second moon. His claim was either ignored or ridiculed. Whereas Waltemath was clearly a crank, the hypothesis of a second moon of the Earth, or a satellite encircling the Moon, was entertained also by a few astronomers.

distance of 427,250 km, that is, moving on the far side of the Moon. The editors of *Science*, to which he reported his discovery, did not take it serious at all. They unceremoniously suggested that the moon presided over lunacy.[49]

The famous Harvard astronomy brothers, Edward Charles and William Henry Pickering, had an interest in the possibility of a second moon, but in a very different sense than the one proposed by Petit and Waltemath. It seems likely that this interest was related to Hall's discovery of 1877 of the two satellites of Mars. In early 1888, Edward C. Pickering, the director of the Harvard College Observatory, wrote to Hall about "looking for additional satellites of other planets [apart from Mars]." He did not mention Venus, but asked: "Would a satellite placed at a greater distance from its planet continue under any circumstances to accompany the planet, or would it become an independent member of the system?"[50] This was the kind of consideration that much later led to the hypothesis that Mercury is a lost satellite of Venus. The two astronomy brothers thought that by means of photography "we may expect to discover such a satellite, if it exists, or to prove that no such object above a certain limit of brightness revolves around the Moon."[51] During the lunar eclipse of 28 January 1888 they took a series of photographs which they carefully examined for a possible satellite, without finding a candidate. They concluded from their negative search that "it is probable that the Moon has no satellite more than 200 metres in diameter."[52]

The failed search did not deter them from further investigations. Edward Pickering observed together with his assistants a total lunar eclipse which occurred on 15 November 1891. At the same time, his brother Henry made observations in the mountains of Peru. Part of the observation programme was aimed to search photographically for a lunar satellite, that is, a small satellite orbiting the Moon. According to the *New York Times*, the astronomers wanted "to confirm Prof. [E. C.] Pickering's opinion that the moon has no satellite."[53] None of the photographic plates, either from Harvard or Peru, revealed a satellite encircling the Moon. Although no lunar satellite was found, the photographic method proved successful: In August 1898 William Pickering found in this way Saturn's ninth moon, which he named Phoebe, the first satellite of any planet to be discovered photographically.

Photography made the search for small celestial bodies more accurate and

[49] *Science* **8** (12 August 1898), p. 185, section on "Scientific notes and news."

[50] Quoted in Gingerich 1978, p. 130.

[51] Pickering and Pickering 1890, p. 83. Further details on the search for a lunar satellite are given in Baum 1973, pp. 19–47.

[52] Pickering and Pickering 1890, p. 83.

[53] "Studying the eclipse; how the Harvard observers viewed the Moon," *New York Times*, 22 November 1891, pp. 17–18. It is not obvious why Edward Pickering (1846–1919) took a stand on the question of a satellite of the Moon. As far as we know, no one had suggested the existence of such a body.

objective, but it was no sure way of avoiding false results. This is illustrated by Pickering's discovery claim in 1905 of the tenth satellite of Saturn, which he named Themis and found in the same way as Phoebe. The following year Pickering was awarded the prestigious Lalande Prize for his "discovery of the ninth and tenth satellites of Saturn," but it soon turned out that Themis was a mistake. No other astronomer was able to confirm Pickering's claim. (Saturn does have a tenth satellite, named Janus, but it was only discovered in 1966 and has nothing to do with Themis.)

Another distinguished American astronomer, Edward Emerson Barnard, was interested in the possible existence of a lunar satellite, although without having much faith in it. From the Lick Observatory he made careful photographic studies during the lunar eclipses occurring on 10 March and 3 September 1895. However, the result was disappointing, if not perhaps surprising: "None of these pictures made during the two eclipses shows anything which might be taken for a lunar satellite." According to Barnard, this was strong evidence against the hypothesis of such a body. He concluded that "a further search for it therefore appears quite unnecessary."[54] In spite of Barnard's conclusion, a few further searches for a satellite of the Moon were made. The most notable of these was conducted by Clyde William Tombaugh, the discoverer of the planet Pluto (or what until recently used to be planet Pluto). Tombaugh's search, which was completed in 1959, was no more successful than earlier searches.

The possibility of a meteoric satellite, the same kind of object that Petit thought to have established in 1846, was considered by William Pickering in one of his many articles in *Popular Astronomy.* By no means foreign to speculation, he was known as an enthusiastic pluralist and for having proposed a number of hypothetical planets – so why not a second satellite of the Earth?[55] Although he did not claim that such a body existed, he believed it was a possibility that should not be ignored. The body would be too small and move too rapidly to leave a trail on a photographic plate, but Pickering suggested that amateur astronomers might chance to detect it with their telescopes.[56] His advice of how to observe the hypothetical satellite caused some amateur astronomers to search for it, but no additional moon was found.

Finally, a German amateur astronomer by the name W. Spill thought that he had seen the second moon in an observation made on 24 May 1926. In a report to the journal *Sirius* he wrote that he had observed a small darkish sphere of apparent size 6" and that it could neither be a balloon nor a meteor. "In 7 seconds the small body passed with a scarcely diminished velocity almost the precise diameter of the

[54] Barnard 1895, p. 347. See also Baum 1973, pp. 33–34 and Sheehan 1995, p. 288.
[55] On Pickering's claims of undiscovered planets, see Hoyt 1976.
[56] Pickering 1923.

Moon (I estimated it to be 200 km west of the Moon's centre); 3 seconds after it had left the Moon's disk it suddenly turned gray like iron". Spill believed he had confirmed what he took to be Pickering's idea of a second satellite: "When I contemplate all the appearances of the observation of 24 May 1926, I feel certain that I, by a happy accident, have been able to observe the second moon of the Earth that professor Pickering has discovered."[57] Although Spill's claim of 1926 was not the last of its kind, and there have been a few proposals of additional natural satellites of the Earth even in the second half of the twentieth century, we shall ignore these and instead return to the moon of Venus.

6.4 Twentieth-century postscripts

Working with the new 36-inch telescope of the Lick Observatory during the summer of 1892, Barnard was especially interested in new satellites. His search for faint Jovian satellites resulted in the discovery on 9 September of the fifth satellite of Jupiter, called Amalthea, the last satellite to be discovered visually.[58] The news of the discovery was announced two days later and created a minor sensation in the astronomical community.

About a month earlier, on 13 August, half an hour before sunrise, Barnard recorded a star-like object near Venus. He estimated it to be 1 degree south of the planet and at least of magnitude 7. The object was not listed in any star catalogue and since it could not be one of the brighter asteroids either, it puzzled Barnard who noted that "There can be no mistake in the date, and a reflection of Venus is out of the question." In his communication of the observation to the *Astronomische Nachrichten*, only published in 1906, he further stated: "The elongation of Venus from the Sun was about 38°which would exclude the possibility that the object was an Intra-Mercurial planet, but it does not preclude the possibility of its being a planet interior to Venus, though such is not probable. ... a reflection from the image of Venus is out of the question."[59] He did not suggest the possibility of a satellite of Venus, although one may imagine that the idea crossed his mind. The most plausible explanation may be that Barnard had observed a nova, a possibility

[57] Although he may have discussed the subject informally, William Pickering (1858–1938) never claimed to have predicted or discovered a second moon of the Earth. Spill's observation report appeared in Kritzinger 1926.

[58] Barnard 1892. The name Amalthea was suggested by Flammarion in correspondence with Barnard.

[59] Barnard 1906. Rudolph Pirovano, an Austrian astronomer, thought to have found an inconsistency in the data given by Barnard. The American astronomer replied by providing further details of his observation of 1892, assuring that his data were correct. See the discussion in *Astronomische Nachrichten* **172** (1906), columns 207–208 and **173** (1907), columns 315–318.

first mentioned by John Ellard Gore, the Irish author on astronomy.[60]

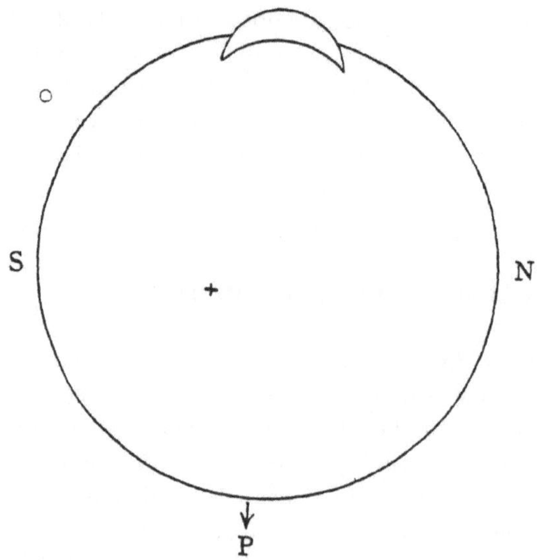

Figure 6.6: Barnard's sketch of his "unexplained observation" of 13 August 1892. The size of the field was 5'. The Venus crescent is seen at the top and the unexplained object is shown beneath it. The sketch dates from the observation of 1892, but was only published in Barnard's paper of 1906.

A few years after Barnard's report, the controversial American astronomer Thomas J. J. See, working at the Naval Observatory at Mare Island, California, published a theory of how the Moon had come into existence. According to See, the Moon had been captured in its orbit by the Earth, and similar capture processes were responsible for the existence of the other satellites of the solar system. He believed that it was "by no means improbable" that the Earth had a small undiscovered satellite beyond the Moon. Whereas his calculations showed a Mercury moon to be highly unlikely, this was not the case for Venus: "Venus holds out better prospects of possible discoveries; for the closed surface about this planet has more than twice the diameter of the orbit of our Moon, and is therefore ample for holding one or more satellites. As Venus admits of prolonged photographic search

[60]Gore 1909, p. 30. On Barnard's observation, see also Sheehan 1995, pp. 200–202 and Baum 1973, pp. 87–91.

when at elongation, this method would be worthy of trial."[61] See did not follow up the remark, nor did anyone else at the time search photographically for a moon orbiting Venus.

The case of Roedkiær's observation of 4 March 1764 as a possible detection of Uranus was re-examined as late as 1963. Jean Meeus, a Belgian astronomer and meteorologist, confirmed Stroobant's conclusion that the object could not possibly be Uranus and argued, in agreement with Hell, that the Copenhagen astronomers had been the victims of optical illusions. According to Meeus' recalculations, on all the dates other than 4 March, Uranus was one or more degrees distant from Venus and thus could not have been the object recorded on these other dates. In addition, the nearness to Venus of the object observed on 4 March, which Roedkiær estimated to be half a Venus diameter, did not fit with the computed position of Uranus relative to Venus at the time of observation. As Meeus pointed out, illusions are no more illusory than they can be detected on a photographic plate and hence acquire a degree of apparent objectivity. Interestingly, he referred to a French astronomer, Roget Rigollet, who on 12 March 1948 "thought to have discovered photographically a satellite of Venus; the negative was obtained at the Paris Observatory with the help of an astrograph of 17 cm opening and 120 cm focal length. It was soon noticed that the 'satellite' revealed a complete steadiness with respect to the planet, what left no doubt as to its nature."[62]

Still in the 1950s the possibility existed that Venus might possess a moon of a kind, only so small and inconspicuous that it escaped even the best telescopes. The concept or name "moon" refers to a body revolving around a planet, but it does not limit the size of the body, neither in absolute nor relative terms. In his book on Venus of 1956, the noted British astronomer Patrick Moore wrote that "it is not impossible that Venus may have a tiny companion," but the sentence was left out in the edition of 1982 coauthored with Gary Hunt.[63] The change was undoubtedly due to the exploration of Venus with artificial satellites, starting in the 1960s with the Soviet Venera probes and the American Mariner probes. In 1978 Venus was even provided with a satellite, albeit an artificial one, when the space probe Pioneer-Venus 1 (also known as Pioneer-Venus Orbiter) was placed in an elliptical orbit around the planet. Although many of the early spacecraft missions to Venus were unsuccessful, by the 1980s the probes had provided astronomers with a new and reliable picture of the planet and its surroundings. With the planet being monitored by satellites constantly transmitting pictures to the Earth, the romance of the natural satellite of Venus finally came to an end.

[61] See 1909, column 345. On See (1886–1962) and his dubious reputation, see Sheehan 2002.

[62] Meeus 1963, pp. 38–39. If Rigollet reported his astrographic observation, we have not found the report.

[63] Moore 1956, p. 96; Hunt and Moore 1982.

Or almost so. One obvious way of making sense of the bewildering appearance and disappearance of the Venus moon would be to postulate that it was there in the past and then, somehow and at some time, disappeared. If the observations were of a real moon, such a knocking-off process would be hard to explain, but what if the satellite were artificial? A few enthusiasts have suggested that what Montaigne and his colleagues saw in the eighteenth century was an unidentified flying object, perhaps a huge extraterrestrial spaceship of some sort, subsequently dismantled or moved to some other corner of the universe. Having recounted the early observations, a Rosicrucian publication of 1958 asked: "Were these observations illusions? Were they real, or has the density of the atmosphere covered them? Is this Venusian atmosphere expanding? Could they have been artificial satellites, and, perchance, are we only 300 years behind the times?"[64] Again, according to a more recent UFO newsletter, "If the objects seen in 1740 and 1764 were merely asteroids drifting through the inner solar system, then they should have been still visible days or weeks later." They were not, which "suggests a powered spacecraft moving out of Venusian orbit."[65] This is a neat if fanciful explanation, but strangely enough it has not won the acclaim of either astronomers or historians of science.

The idea to explain away the sightings of Venus' moon by identifying it with a spacecraft may seem outrageous, and of course astronomers do not take it seriously. But it is scarcely more audacious than a somewhat similar suggestion made in 1959 by the brilliant Russian astrophysicist Iosef Shklovskii in connection with Mars' inner moon, Phobos. A pioneer radio astronomer and the recipient of the prestigious Bruce Medal, Shklovskii was a highly reputed scientist; more controversially, he was also a pioneer of modern SETI research (Search for Extraterrestrial Intelligence). In order to explain why the orbit of Phobos was apparently shrinking, he suggested that the upper regions of the Martian atmosphere might gradually slow down the satellite in its orbit about the planet. Calculations proved that this would work if Phobos were assumed to be hollow, which implied that it was a technological product – in effect a gigantic artificial satellite "on a scale surpassing the fondest dreams of contemporary rocket engineers." Shklovskii admitted that the idea might seem fantastic, but he nonetheless thought that astronomers should seriously consider the idea that "Phobos was launched into orbit in the heyday of a technical civilization on Mars, some hundreds of millions of years ago."[66] He did

[64] Burbidge 1958, p. 168.

[65] *UFO Roundup*, 5 October 2000 (`http://www.ufoinfo.com/roundup/v05/rnd05_40.shtml`). See also Moore 1956, p. 96, and Corliss 1979, pp. 137–139. Several observations of unidentified shining objects near Venus are reported in
`http://www.xdream.freeserve.co.uk/UFOBase/Astronomers.htm`.

[66] Shklovskii and Sagan 1966, p. 373–374. The book was a revised translation of a Russian work of 1962, with Shklovskii (1916–85) as the sole author. Later in life, he abandoned his belief

not believe that Martians, technologically advanced or not, were still alive.

On a more scientific note, scenarios of how Venus might have lost its moon at a stage in the early history of the solar system have been discussed by planetary scientists since the 1970s. It is possible that Venus, and also Mercury, originally had a satellite but that they lost it because of tidal friction that caused it to recede.[67] According to some calculations, at some stage the effect reversed and the satellite was drawn in again and crashed on the planet. As late as in the fall of 2006 David Stevenson and Alex Alemi from the California Institute of Technology suggested that the mystery of the missing moon might be explained by assuming two large impacts: The first giant impact created a moon and caused Venus to spin counterclockwise (as seen from above the planet's north pole); the second impact, taking place a couple of million years later, reversed the direction of rotation and caused the moon to move inwards and eventually collide with the planet.[68] Whatever the scientific merit of this and other theories, should Venus once have had a moon, it was long before there were humans to observe it. Consequently these theories, interesting as they are from the point of view of astronomy, have no real significance for the history of the satellite of Venus such as it evolved during the period from the late seventeenth century to the late nineteenth century.

As the satellite of Venus entered the literary world of the Enlightenment with such works as Bonnet's *Contemplation de la nature* and Martin's *Young Gentleman and Lady's Philosophy*, so it entered into the literature of the early twentieth century. But it did so in very different kinds of works, less reputable than those of the eighteenth century. For example, the long discarded moon continued to attract attention in some esoteric and astrological circles. The leading clairvoyant and theosophical author, Charles Webster Leadbeater, a close friend of Annie Besant, referred in 1911 to the "remarkable fact that astronomers of a hundred and fifty years ago recorded several observations of a satellite of Venus." Although aware of the alleged moon, apparently he had not bothered to check its history: "It was seen by astronomers as well known as Cassini and Short, in 1761, and that not one but many times, and again in 1764 by Rödkier, Horrebow and Montbaron." Unable to believe that the observations were nothing but errors, Leadbeater suggested that they agreed with theosophical doctrines, "for we are told that in our seventh round the moon will disintegrate and we shall be left without a satellite; it may only be a coincidence that Venus is in its seventh round, but it is a curious one."[69]

in extraterrestrials and turned towards anti-pluralism.

[67]See Burns 1973 and Ward and Reid 1973. Other astronomers have suggested the possibility that Mercury might once have been a satellite of Venus, but that tidal interactions caused Mercury to escape into a solar orbit. This line of research started with Van Flandern and Harrington 1976.

[68]Alemi and Stevenson 2006. See also *Scientific American*, online edition, 10 October 2006.

[69]C. W. Leadbeater, *The Inner Life: Theosophical Talks at Adyar* (Chicago: Rajput Press,

The controversial American author Charles Hoyt Fort is known in particular for *The Book of the Damned*, a work published in 1919. Fort felt strongly attracted by phenomena that were paranormal or otherwise went beyond established science, and he collected a large amount of such "damned data" rejected by what he saw as the scientific priesthood. Among them were the satellite of Venus and also the planet Neith, the rejection of which he associated with what he considered to be the "dogmatism and pontification" of science. Unsurprisingly, he also expressed sympathy with the second moon of the Earth, supposedly discovered by Petit. "We see only what it is 'proper' that we should see," Fort complained. "I think there would not be much approximation to realness in taking refuge in the notion of astronomers who stare and squint and see only that which it is respectable and respectful to see." The long discarded Venus moon was for him an example of "celestial vagabonds [which] have been excluded by astronomers, primarily because their irresponsibilities are an affront to the pure and the precise, or to attempted positivism; and secondarily because they have not been seen very often." However, these celestial vagabonds

> have been seen and reported so often that the only important reason for their exclusion is – that they don't fit in. ... What I emphasize here is that our damned data are observations of the highest standing, excommunicated by astronomers of similar standing – but backed up by the dominant spirit of their era – to which all minds had to equilibrate or be negligible, unheard, submerged. ... In 1645, a body large enough to look like a satellite was seen near Venus. Four times in the first half of the 18th century, a similar observation was reported. The last report occurred in 1767. A large body has been seen – seven times, according to Science Gossip, 1886–178 – near Venus.[70]

Had Paul Feyerabend, the philosopher of epistemic anarchism, known about the case of the satellite of Venus, he might have agreed. We are not aware of social constructivists who have argued that the satellite exists – socially constructed, that is – but why not?[71]

1911), here quoted from the online edition `http://www.anandgholap.net/Inner_Life_Vol_II-CWL.htm`.

[70] Fort 1999, p. 184, here from a hypertext edition (`http://www.resologist.net/damn14.htm`). On Fort and Forteanism, see Gardner 1957, pp. 42–54. Constance 1956 is another book in the strange-object tradition of Hoyt and Corliss. It deals with the Venus satellite on pp. 45–46.

[71] For an example of social constructivists' implicit defense of discarded entities, in this case the notorious *N*-rays, see Ashmore 1993. And for critical comments, Kragh 1998.

Chapter 7

Conclusion, and a note on the satellites of Uranus

The case of Venus' phantom satellite has traditionally been considered a mere curiosity in the annals of astronomy. A curiosity it may have been, but it was much more than that. As the present study demonstrates, the hypothesis – and it never was more than that – played a considerable role in the eighteenth century, primarily among astronomers but also in a wider context. As illustrated by figures such as Bonnet, Voltaire, d'Alembert, Kant, Herder, Martin and Frederick II of Prussia, the cultural world was acquainted with the phenomenon and found it to be of interest. At least on two occasions, enlightenment scientists (A. G. Kästner and L. A. Jungnitz) even wrote poems dedicated to the controversial satellite. From the beginning of the story, with Fontana's observations in 1645, to the late nineteenth century, there was a rich literature on this non-existing object, and it was much richer than has traditionally been thought. Admittedly, much of this literature was repetitious, but it nonetheless indicates an interest in, and to some extent a fascination with, the satellite of Venus.

Fontana's observation of a companion to Venus was well known in the seventeenth century, but it was not considered more reliable than the contemporary observations (by Rheita and Fontana) of five additional moons around Jupiter. What distinguishes the two cases of spurious discovery is that the satellite of Venus was seen twice by Cassini and later by Short and several other observers. Although observed many times, the enigmatic satellite never made it to become an accepted part of astronomy. In fact, only few of those who observed it claimed explicitly that what they had seen was a satellite of Venus (namely, Scheuten, Roedkiær and Horrebow, and possibly Montaigne). Also Mairan and Baudouin were convinced of its existence, at least for a time, but none of them had actually seen the moon or the image that looked like it. Neither had Lalande, and yet he was for a period of more than twenty years sympathetically inclined towards the satellite.

The relatively few discovery claims were not accepted by the astronomical community, which found the evidence for them to be wanting in quality and, not least, in quantity and regularity. For this reason alone, one cannot speak of a discovery, not even of a spurious object. A further indication of the undecided status of the Venus moon is the absence of eponymy – a name was never proposed for it (King Frederick's suggestion of "d'Alembert's satellite" was informal and merely an expression of flattery). Horrebow firmly believed that he and his assistant had discovered the satellite, yet he refrained from affixing a name to it. Whereas Houzeau was quick – indeed too quick – to announce his hypothetical planet as Neith, Schorr, writing at about the same time, refrained from suggesting a name for the moon in which he clearly and heterodoxically believed.

In spite of the general agreement that Venus did not possess a moon, something imitating a moon *was* seen in the telescopes, and from the mid-1760s onwards the problem became to account for the observations without taking the satellite hypothesis too seriously. This problem was routinely addressed in the astronomical literature. It was considered by several astronomers, from Hell to Stroobant, who over more than a century came up with a number of suggestions. The more important of these explanations of why something had been mistaken for the satellite of Venus were the following:

(i) Optical illusions produced in telescope and eye, i.e. Hell's explanation.

(ii) Faint stars mistaken for the moon, which was Stroobant's favourite explanation.

(iii) A new interior planet, as proposed by Houzeau.

(iv) Asteroids, such as first suggested by von Ende.

(v) Uranus, being mistaken for the satellite, such as first proposed by Jean Bernoulli in 1781.

(vi) Atmospheric illusions caused by diffraction in ice crystals, as proposed by Thirion in 1884.

(vii) Sunspots mistaken for a satellite.

In addition, Bianchini, Jacques Cassini and Mairan suggested in the eighteenth century that the observations might be due to condensations of the fluid matter that supposedly filled up the space of the solar system. However, this hypothesis fell with the abandonment of the interplanetary fluid matter in the wake of the transition to Newtonian physics. Of the suggestions listed above, (iii) and (iv) were not taken very seriously – they explained observations by introducing an unconfirmed hypothesis. Nor was explanation (v) widely accepted, if for no other reason that at best it could explain only a few of the observations. Explanation (vi) may have come too late to attract much interest, and explanation (vii) was only relevant in the few cases when a moon-like object had been seen during a Venus transit. The most popular explanations were clearly (i) and (ii), which were often used in combination – if one of the explanations failed in some cases, the other might be appealed to.

At any rate, after Stroobant's work in 1887 astronomers lost interest in the matter, confident as they were that the mystery had been solved. It continued to be reported in the more popular literature, typically as an instructive case of how errors and illusions might enter astronomical observations.[1] The account in the 1911 edition of the *Encyclopædia Britannica* was quite typical. After briefly having mentioned the observations of Cassini, Short and others, it said: "But as no such object has been seen by the most careful search with the best instruments of recent times, the supposed object must be regarded as what is known to the practical astronomer as a 'ghost' produced by reflection from the lenses of the eyepiece, or perhaps of the object-glass, of the telescope." We have not tried to determine what really was seen in the various observations, that is, to determine whether Mairan, Hell, Stroobant, Thirion or other post-1760 commentators were right or not; nor do we believe that such an undertaking would be historically fruitful (it may be of a certain astronomical interest, but that is a different matter).

The history of science amply illustrates that "believing is seeing," in so far that it includes numerous examples of scientists who have seen things that do not exist, either for psychological, social or instrumental reasons. Among those examples, the N-rays that were discovered by the French physicist René Blondlot and investigated in the early years of the twentieth century is a classical case that has received a great deal of attention from historians of science.[2] Many false observations have been rooted in or stimulated by preconceptions, more or less articulated beliefs that something should exist and therefore do exist. Such beliefs or expectations may be theoretically grounded, but in other cases they are based on vague analogies or sheer wishful thinking.

To mention but a few examples from the history of planetary astronomy, Thomas Harriot's early drawings of the Moon from 1609–10 included features that almost certainly had their origin in his mind rather than in his telescope. As mentioned in section 2.2, Bianchini thought in 1728 that he had identified

[1]For a couple of examples from about the turn of the century, see Plassmann 1898, p. 355, Krisch 1901, p. 532 and Vogel 1905, pp. 345–346.

[2]Nye 1980; Ashmore 1993.

continents and oceans on the surface of Venus. At the end of the eighteenth century, Schröter believed that he had seen very high mountains on the planet and reported details that he could not possibly have seen. More than a century later, Percival Lowell reported that he had seen clearly defined markings on Venus – other astronomers failed to verify the observation. The most well known case of deception is undoubtedly Lowell's sensational discovery claim of canals on Mars and the furor it created.[3]

Yet another example is provided by the great William Herschel, who on 13 March 1781 observed a puzzling object which he initially took for a comet. He was able to measure its apparent diameter and, in conformity with his expectations, found that it "certainly increased, from which we may conclude that the Comet is approaching to us." In reality, the size of the object – the planet Uranus – decreased during the period of observation. It was only in mid-April that Herschel began to realize that it was not a cometary object, but something else; and it took another year until he referred to it as a new planet.[4] Again, in 1846–47 the reputed British amateur astronomer William Lassell observed a ring around the recently discovered planet Neptune, although what he saw did not actually exist.[5] His expectation that Neptune was accompanied by a ring influenced his observations. Interestingly, Lassell was not the only observer to see Neptune's non-existing ring. So did a few others, including James Challis, the Plumian Professor of Astronomy in Cambridge, who was aware of Lassell's claim. The more or less independent observations made the Council of the Royal Astronomical Society declare in early 1847 that "the existence of the ring seems almost certain."[6]

Telescopic observations are often delicate and may blur the line between what is actually observed and what the observer wants to see. Barnard had found Jupiter's fifth satellite (Amalthea) on 9 September 1892, but for some time the faint object eluded European astronomers. E. Walter Maunder at the Greenwich Observatory was among those who much wanted to see the satellite. In early 1893, after several failed searches, he wrote to Barnard: "I have tried hard again and again to catch a glimpse of your fifth satellite with our new 28-inch telescope, but only succeeded on two occasions in just *fancying* I saw it for a moment."[7] He was

[3]On Lowell and his disputed observations of Venus and Mars, see Sheehan 1988.

[4]See Austin 1967 and Schaffer 1981.

[5]For this case, see Baum 1973, pp. 126–146 and Smith and Baum 1984. Neptune does have a system of very thin rings, but this was only discovered by the Voyager spacecraft in 1989. Neither Lassell (1799–1880) nor other Earth-bound observers could possibly have seen it. Hetherington 1988 and Sheehan 1988 offer several other examples of "believing is seeing" in the history of astronomy. On the general problem of objectivity versus visual perception and representation, see Daston and Galison 2007 and the many sources cited therein.

[6]Cited in Baum 1973, p. 141.

[7]Sheehan 1995, p. 215. European observers were also unable to see Mars' inner satellite, discovered in August 1877; it took about half a year until they succeeded in verifying Hall's

well aware of the danger of substituting observation with the fancy of observation.

Instructive as these cases are, the story of Venus' moon belongs to a different category, in so far that preconceptions played no or only a very limited role. Most of the astronomers who thought they had seen the moon had no preconceived convictions that Venus ought to have a companion. They just happened to see something in their telescopes that might or might not be a satellite. This was a reasonable hypothesis, but it was no more than that. Theory did not support it, nor did it speak against it. In fact, if there were a preconception with respect to Venus' satellite, it was that there probably was no such satellite, and that for the reason that Venus is an interior planet. As we have seen, it was widely believed that the number of satellites increased with the distance from the Sun, the reason being that the outer planets were in more "need" of moons.

Scientists of the eighteenth century were well aware of the problems of observation and the difference between scientific observation and brute visual perception. Such problems were evident to those scientists who relied on instruments such as the microscope or the telescope. The Geneva clergyman and physiologist Jean Senebier published in 1775 a pioneering account of these and related problems, the *L'art d'observer.* He noted that rare phenomena were particularly puzzling and referred in this connection to the halos and meteors known from meteorology and astronomy. Moreover: "Astronomy is interested in the observations of the zodiacal light and the satellite of Venus, whether true or supposed."[8]

The reluctance of late-eighteenth-century astronomers to admit a satellite of Venus should be seen on the background of the conservative climate with respect to new celestial bodies – comets excepted – in the solar system. Ever since the victory of the Copernican system there had only been six planets (rather than seven), a number which initially was seen as pleasing because it is the simplest perfect number – this was an argument employed by Rheticus, Copernicus' only pupil, and it was also used by Huygens.[9] Although the magic of perfect numbers soon evaporated, when Uranus was discovered in 1781 it came as a complete surprise. It is well known that Herschel initially resisted classifying his discovery as a planet.[10] Even after the number of planets had increased from six to seven,

observation (Hall 1878, p. 208).

[8]Senebier 1802 (second edition), p. 136.

[9]In his *Narratio Prima* of 1540, the first published work on the Copernican world system, Georg Joachim Rheticus (1514–1574) reasoned: "What is more agreeable to God's handiwork than that this first and most perfect work should be summed up in this first and most perfect number?" Rosen 1959, p. 147. Kepler adopted Rheticus' reasoning (Koyré 1961, p. 139). On Huygens' use of a similar argument, see section 2.1.

[10]As mentioned, Herschel discovered Uranus in 1781 but at first thought it was a comet (Austin 1967 and Schaffer 1981). On his discovery of the satellites of Uranus, see Lubbock 1933, pp. 161–168.

conservatism remained, such as shown by Giuseppe Piazzi's reluctance to acknowledge his finding in early 1801 of an asteroid (Ceres) as a new member of the solar system.

The conservatism may not have been quite as strong with regard to the secondary planets, but after 1684 most astronomers took it for granted that the number of satellites had stabilized to ten. Although there were no good reasons why still more satellites (or, for that matter, planets) should not exist, neither were there any good reasons to expect they did exist. Most astronomers considered this kind of argument to be persuasive. When it comes to ontology, science is essentially conservative. If scientists have to consider an increase of natural objects, and there are no strong theoretical or empirical reasons for such an increase, they will typically regard it as "uneconomical" and therefore oppose it. They will tend to adopt an ontological version of Ockham's famous razor. By the 1760s the number of satellites had remained unchanged for about eighty years, and to accept new ones would require strong observational evidence. As the majority of astronomers saw it, in the case of Venus' moon such evidence was lacking. The attitude with respect to the moon of Mars, occasionally discussed in the literature, was the same.

The quick change in attitude during the 1760s and 1770s presents something of a puzzle. After the last public discovery claim of 1764 it took less than a decade until the credibility of the Venus moon was eroded. Obviously the discovery claim was not proved wrong in the ordinary sense, nor could it be. Because, how does one prove that a celestial body does not exist? The claim that there exists a satellite of Venus is an example of what Karl Popper calls a restricted existential statement, one which says that something exists under certain specified circumstances.[11] According to Popper, whereas a strict existential statement of the type "X exists" is irrefutable (and hence non-scientific), this is not necessarily the case with statements of the restricted type. The claim that Venus possesses a moon is obviously verifiable, but until the advent of spacecraft in the 1960s it was in practice non-falsifiable. Yet, the astronomers, unconcerned with Popperian philosophy, considered it a question that could be solved – and in fact was solved – by scientific means.

The satellite of Venus was "de-discovered," but only passively, in the sense that after 1764 no further observations were reported. The mere absence of confirmation counted as a refutation. This may be compared with Cassini's observation, which remained unconfirmed for more than fifty years and still enjoyed fairly wide (if not general) recognition, undoubtedly in part because of the high reputation of the distinguished Italian-French astronomer.

More generally, one may ask: If an observation is not confirmed, what does

[11] Popper 1972, pp. 195–197.

it take to disconfirm it? How many years have to pass until lack of confirming evidence counts as a disproof? Or, on an even more general level, what is the relationship between lack of evidence and counter-evidence? This is far from a trivial question, and it is not one that philosophers of science have paid a great deal of attention to. The formula that "absence of evidence is not evidence of absence" is generally accepted, but not very helpful. Logic apart, in the real world of science it often happens that lack of corrobative evidence counts as evidence of absence. Venus' satellite was not really a case of "absence of evidence," for several astronomers had seen the object; it was rather a case of absence of convincing evidence where the many non-observations amounted to evidence of absence. The many observations in connection with the transits of 1761 and 1769 that failed to reveal the attendant of Venus are undoubtedly part of the explanation of why the astronomers declared it non-existent, but they do not explain it fully, nor do they explain why it happened so quickly and decisively. After all, there was no more reason to believe that Venus was companionless than Mars was without a satellite or – relating to the years between 1781 and 1787 – that the new planet eventually called Uranus possessed no moon.

It is instructive to compare the reception of the Venus moon with the corresponding case of the satellites of Uranus. William Herschel was convinced that what he called the Georgian planet had several satellites, mainly because it was heavy and so far from the Sun. In 1787 he reported to have discovered two of the moons, later called Titania and Oberon. Although he did not see them very clearly, he was able to follow the moons for as long as nearly five hours, considerably longer than the Venus moon was ever seen in a telescope. It may be a matter of discussion whether the evidence for Titania and Oberon was solid or not, but what matters is that neither Herschel nor others had any doubts about their existence. The two moons became facts from the very beginning. Then, at the end of the 1790s, he announced the discovery of four additional satellites.[12]

As far as we can tell, no one suggested that the Herschelian moons were telescopic phantoms, due to faint stars, concentrations in the ether or otherwise mistakes. For example, the distinguished French amateur astronomer and mathematician, Achille-Pierre Dionis du Séjour, stated in a work of 1789 that Uranus had two moons, presenting it as if it was a plain matter of fact. Contrarily, as to Venus, he said in a likewise matter-of-fact way that "this planet has no satellite." The following year the German scientist Ernst Gottfried Fischer opined that the Uranus moons were now incontestable facts – "truths, no longer hypotheses."[13]

[12]On Herschel's work on the satellites of Uranus, see Herschel 1912, vol. 2, pp. 1–21 and 542–574. See also Schorr 1875, pp. 52–60 and Alexander 1965, pp. 58–76.

[13]Dionis du Séjour 1789, p. 531, who referred to the planet as "Herschel," a name proposed by Lalande. E. G. Fischer (1754–1831) suggested on a speculative basis – the longer away, the

Also the French astronomical writer M. Voiron, author of a history of astronomy between 1781 and 1811, described Herschel's discoveries of the Uranus moons as if they were unproblematic; and this not only for the two satellites of 1787 but also for the four more that Herschel claimed to have found by 1798.[14] Likewise, in 1816 Bode described the six moons as if they were commonly accepted celestial bodies.[15] As far as we are aware, no astronomer in the period from 1787 to the 1830's expressed serious doubts as to whether Herschel's six satellites of Uranus did really exist.

Herschel continued his series of observations of the Uranus moons until 1810, but in spite of his dogged efforts he only succeeded in seeing the suspected moons a very few times and at widely scattered dates. For example, on 18 January 1790 he observed: "A supposed 3d satellite is about 2 diameters of the planet following; exceedingly faint, and only seen by glimpses." And on 14 March, the same year: "Supposed 3d and 4th satellites were observed, but no opportunity be had afterward to see them again."[16] In no case was he able to observe any of the objects on consecutive dates. Nevertheless, based on such questionable evidence he concluded that the moons were real. Clearly – at least from a modern perspective – the evidence for the satellites was slender, and this was the case especially for the four satellites announced in 1798.

The less than solid evidence is underscored by the near lack of confirmation from other astronomers using telescopes different from the one used by Herschel in his "Observatory House" in Slough. The only other astronomers who claimed to have seen the two satellites reported in 1787 were Schröter and his able assistant Karl Ludwig Harding, who in February 1797 succeeded in observing Oberon and Titania on three consecutive nights.[17] Apart from this observation, William Herschel had what essentially was a monopoly on the Uranus satellites until the late 1820s, when his son, John Herschel, was able to confirm the existence of the first two moons. Although the younger Herschel found no evidence of the other satellites, yet he expressed no doubts about their existence and hoped soon to find

---

more satellites do the planets have – that Uranus was endowed with six moons (Fischer 1787).

[14] Voiron 1810, pp. 22–25. Also Johann Friedrich Wurm (1760–1833), a professor of astronomy at Stuttgart, referred uncritically to Herschel's six moons, apparently accepting them (Wurm 1802). Wurm had earlier speculated that the known satellites of Jupiter and Saturn were represented by numerical expressions of the same kind as the Titius-Bode law (see Nieto 1972, pp. 24–25).

[15] Bode 1816, p. 343.

[16] Herschel 1912, vol. 2, p. 9 and p. 4.

[17] Schröter 1798. Strangely, John Herschel (1792–1871) seems to have been unaware of Schröter's confirmation. In 1834 he wrote about the satellites of Uranus that, "they have never been observed, or even seen (as far as the author is aware), except in the telescope in which they were originally discovered." Herschel 1833–36, p. 35.

them.[18] The only "confirmation" of these satellites came in 1837, when Johann Lamont, director of the Munich Observatory, saw the first two of Herschel's satellites and also, if only on one occasion, what he thought was the most distant of the supplementary satellites. Lamont realized that his observation was not very reliable and consequently left it out in his calculation of the mass of Uranus.[19]

To summarize a long and complex story, it is remarkable that for about half a century almost all astronomers accepted the existence of the moons of Uranus, and that in spite of the meagre evidence and almost total lack of independent observation. According to William Smyth, writing in 1844: "On account of the extreme delicacy of these objects, and because for a few years [*sic*] nobody else had succeeded in fishing them up, several continental astronomers were pleased to doubt whether Uranus had any satellites at all."[20] In fact, such doubt scarcely existed and may well have been a product of Smyth's chauvinistic imagination. As to Smyth, he accepted on Herschel's authority that Uranus was surrounded by six moons. And he was not the only one.

Only in 1847 did Lassell and, independently, Otto Wilhelm Struve at the Pulkovo Observatory, see Titania and Oberon as well as one or two of the supplementary satellites. Matters rested there until the fall of 1851 when Lassell finally provided unambiguous evidence for the existence of the two satellites of Uranus that are now named Ariel and Umbriel.[21] However, as Lassell realized, none of these celestial bodies are identical to the Herschelian moons. Both Lassell and Struve later observed all four of the satellites of Uranus extensively. As to Herschel's two last satellites, they vanished from history into the thin air of oblivion. It is now believed that the four supplementary satellites observed by Herschel in the 1790s were in reality very faint fixed stars. (There is a small fifth moon revolving around Uranus, called Miranda, but it was only discovered by Gerard Kuiper in 1948; later several more moons were detected.)

So persuasive was Herschel's reputation that his discoveries were still defended by 1870, many years after Lassell had assumedly clarified the matter. Richard Proctor thought that Uranus had at least eight moons, the six discovered by Herschel *and* the two found by Lassell. The four moons announced in the 1790s "have not been *yet* identified; but one cannot read the account of his method of procedure without feeling that no amount of mere negative evidence can be opposed effectively to the positive information he has left respecting these

[18]Herschel 1833–36.

[19]Lamont 1838.

[20]Smyth 1844, vol. 1, p. 207. According to Schorr, "Several astronomers seriously doubted the existence of the four satellites that had been claimed in the treatise of 1790," but he gave neither names nor references (Schorr 1875, p. 54).

[21]Lassell 1852, letters of 13 and 29 November 1851.

four orbs."[22] This is an interesting statement, of relevance also to the history of the satellite of Venus. Are there cases where isolated positive evidence takes priority over *any* amount of negative evidence?

The relevance of the story of the satellites of Uranus for the case of Venus' moon should be obvious. The case for the six moons of Uranus, say about 1820, was not much stronger than the case for Venus' moon had been fifty years earlier, and yet the astronomical community responded very differently in the two cases. Whatever the reasons, there is no doubt that Herschel's reputation counted very highly – even more highly than Cassini's in the case of the Venus moon. It is also worth pointing out that Herschel's discoveries of moons around Uranus did not prompt a reconsideration of the satellite of Venus. Nor was this the case when Lassell discovered a moon (Triton) around Neptune in 1846 and five years later established the existence of a third and fourth Uranus moon. Again, the discovery of Mars' two moons in 1877 did not in the slightest disturb the conviction of the astronomers that Venus is a lone planet.

It is as if the astronomical community had decided that the companionless Cytherean planet was a fact rather than a convincing hypothesis. And facts are not to be questioned.

[22] Proctor 1896, p. 201. Emphasis added.

Biographical sketches

Baudouin de Guèmadeuc, Armand Henri (1734–1817)

Baudouin worked as a civil servant in Paris, but little is known of his life except that he had an interest in both astronomy and literary culture. At one time he was imprisoned for fifteen months, and later he was forced to leave France. In 1782 he published in Neuchâtel, Switzerland, a collection of anecdotes entitled *L'Espion dévalisé*. The book appeared anonymously and was for a long time ascribed Honoré Mirabaud.

In the spring of 1761 he looked in vain for the satellite of Venus, after which he established contact with Montaigne and requested that he proceeded in the hunt. After Montaigne's success, Baudouin read two papers on the Venus moon to the Academy in Paris. His published version, *Mémoire sur la découverte du satellite de Vénus*, appeared in 1761 and was translated into German and English. Confident that the elusive satellite had now been discovered, he used Montaigne's observations to determine the mass of Venus, which he stated was almost the same as the mass of he Earth. He also predicted the appearance of the satellite on the day of the transit. In company with Messier, he observed the transit on 6 June 1761, but saw no moon. In 1768 Baudouin made observations of Venus and the satellites of Jupiter, which he communicated to the Académie des Sciences. After that time, he seems to have disappeared from the scene of astronomy.

Bianchini, Francesco (1662–1729)

During most of his career, Bianchini served the Catholic church. He wrote a world history (*Istoria universale*) and spent great efforts to improve the accuracy of the calendar. In 1703 he was elected secretary for the commission for the reform of the calendar. As a papal envoy and esteemed scholar, in 1712 he was sent to France and England, where he was received with great respect. In Paris he met with the aging Cassini, and in London he was welcomed by Newton with whom he had several conversations. His best known astronomical work, entitled *Hesperi et phosphori*, was published in Rome in 1728. In this first monograph about Venus, he studied the elusive markings of the planet and determined its rotation period to 24 days 8 hours. He drew a map of Venus and suggested names for several "oceans" and "continents." Craters on Mars and Venus are now named in Bianchini's honour. Although he did not see, nor believe in, the Venus moon, nor did he believe that the observations of Fontana and Cassini were caused by optical illusions. As an alternative he suggested that they were due to a fluid substance in interplanetary space.

Bonnet, Charles (1720–1793)

Born in Geneva, he spent his entire life in Switzerland as an independent naturalist and author. Within natural history, he made several important discoveries, in particular that certain tiny insects could reproduce by means of parthenogenesis. He was in favour of preformation, the view that every creature existed in a preformed state within the egg, and in palaeontology he supported the philosophy of catastrophism. A great advocate of the principle of plenitude, he expounded this idea in his popular *Contemplation de la nature* of 1764. The work dealt briefly with the planetary system and included comments on the satellite of Venus, which Bonnet thought might well exist. There is another connection to astronomy, namely by way of the German translation made by Johann Daniel Titius in 1766. Titius inserted his own speculation concerning the distances of the planets, the first version of what came to be known as the Titius-Bode law. Possibly inspired by Bonnet, the Dutch minister and writer J. F. Martinet discussed the satellite of Venus in his *Katechismus der Natuur* of 1779.

Boscovich, Roger Joseph (1711–1787)

Rudjer Josip Bošković, as his name is also spelled, was born in Ragusa (now Dubrovnik) in Croatia. He entered the Jesuit order in 1725 and began studies at the Collegio Romano in Rome, where he was made a professor of mathematics in 1740. He later became a professor in Pavia and helped (with Louis Lagrange) to establish the observatory in Brera. Throughout his career, he travelled widely and was very well connected. As a correspondent for the Royal Society, he was involved in the preparations to observe the Venus transits of 1761 and 1769. Boscovich is today best known for his dynamical theory of matter, an atomic theory based solely on a universal law of force, which he developed in his *Theoria philosophiae naturalis* of 1758. He also did important work in astronomy and optics, in particular in the theory of telescopes and other optical instruments. Much of this work was collected in a book of 1767, *Dissertationes quinque ad dioptricam*, which included a section on secondary images and the apparent satellite of Venus. Boscovich did not believe in the satellite and explained the observations as optical illusions, much in the same way that his fellow Jesuit Maximilian Hell had done the previous year. Strangely, he did not refer to Hell's investigation.

Cassini, Jean Dominique (1625–1712)

The scientific career of Gian or Giovanni Domenico Cassini, as his Italian name was, began in Bologna, where he studied under Riccioli and Grimaldi. Through patronage connections, in 1650 he was appointed professor of astronomy at the University of Bologna, where he stayed for nearly twenty years. In 1669, he became a member of the Académie Royale des Sciences and was recruited as the first director of the new observatory in Paris. He became naturalized as a French citizen in 1673. During his period in Bologna, Cassini made a series of important observations, which included the rotation periods of Mars and Jupiter. He also came up with a rotation period of Venus a little less than 24 hours. In 1668 he published new tables of Jupiter's satellites, the *Ephemerides Bononsienses Mediceorum Syderum*. In Paris, working with telescopes constructed by Campani, he found two new moons of Saturn in 1671–72 and another two moons in 1684. In 1675 he observed the gap in Saturn's ring system known today as the Cassini division. He was also a pioneer in the study of the zodiacal light, which he thought was of cosmic origin. In 1672 and again in 1686 he saw a faint object near Venus which had the appearance of a satellite, but he did not identify it as such.

Although a brilliant observer, Cassini was conservative when it came to theory. He did not accept the determination of Ole Rømer of the velocity of light and opposed Newton's theory of gravitation. Contrary to the Newtonians, he maintained that the shape of the Earth was a prolate spheroid. During the last years of his life, he was blind. He was succeeded as director of the Paris Observatory by his son Jacques (1677–1756), and also his grandson César-François (1714–1784) and his great-grandson Jean Dominique (1748–1848) became directors of the institution.

Dick, Thomas (1774–1857)

Christian philosopher and high pluralist, the Irishman Thomas Dick was a successful author of popular and morally uplifting books on science. After studies at Edinburgh University, he was licensed to preach, but preferred instead a career as teacher and writer. His first work, The *Christian Philosopher* of 1817, employed natural theology and pluralism to emphasize the divine nature of the heavens. A later book, the *Celestial Scenery* of 1838, included a section on Venus' moon in which Dick surveyed its history and suggested from reasons of analogy that the satellite was probably real. He also considered it probable that Mars might have a satellite.

Fontana, Francesco (ca. 1585–1656)

Known as a telescope maker and observational astronomer, Fontana graduated in law from the University of Naples, but he subsequently devoted himself to the construction of telescopes and other optical devices. He claimed to have built and used a telescope as early as 1608, before Galileo. Although appreciated as a telescope maker, his contemporaries did not hold him in high esteem as an astronomer. Among his many observations were the belts of Jupiter and the phases of Venus. He also observed the 1645 Mercury transit and made a number of drawings of the Moon. These he published in his only printed work, the *Novae coelestium terrestriumque observationes*, which appeared in Naples in 1646. In this work he also reported markings on Venus and one or two small globes observed from November 1645 to January 1646. Although he did not explicitly identify his discovery with satellites of Venus, most later astronomers thought he did. His discovery was mentioned by Riccioli, Gassendi and Kircher, none of whom believed in a Venus satellite. On the other hand, Guericke apparently supported Fontana's claim.

Hell, Maximilian (1720–1792)

Born in Slovakia as the son of a Hungarian family, Hell's original name was Miksa Höll. He entered the Jesuit order in 1738 and after studies of mathematics, astronomy and theology in Vienna he was ordained a priest in 1752. Three years later he was appointed professor of mechanics at the university and the following year the first director of its observatory. Among his many duties was to edit the *Ephemerides astronomicae ad meridianem Vindobonensem* published between 1757 and 1791. He was involved in the 1761 Venus transit project, and in connection with the transit of 1769 he was invited by the king of Denmark to make measurements in Vardø in northern Norway. His data were important, but they arrived late and Lalande thought he had manipulated them, which greatly tarnished Hell's reputation. He was only rehabilitated in 1883, as a result of Simon Newcomb's careful detective work. Hell had an interest in Venus and its alleged satellite, for which he had looked in vain in 1764. But he came to the conclusion that the satellite was spurious and that the observation claims were all due to optical reflections in the telescope and the observer's eye. Based on a series of optical experiments, he argued for this conclusion in *De satellite Veneris*, included in the Vienna Ephemerides of 1766 and translated into German in 1792. Many astronomers found his explanation to be satisfactory.

Horrebow, Christian (1718–1776)

The son of Peder Horrebow, professor of astronomy at Copenhagen University, he followed his father as professor and director of the observatory. Since his election in 1747, he was an active member of the Royal Danish Academy of Sciences and Letters. He managed to improve the instruments and routines at the observatory, but his own observations were not of great importance. Much of his scientific work focused on sunspots, and he was possibly the first to note the periodical variations of the spots. The observations of the Venus transit in 1761, made with Peder Roedkiær and his brother Peder Horrebow, were not of high quality, and he played only a minor role in the 1769 transit observations. In March 1764, he and Roedkiær observed Venus and its supposed moon, and he published an article on the issue in the proceedings of the Royal Danish Academy. Written in Danish, it was not much noticed, but it was known to Hell, from whom other astronomers learned about it. In January 1768, Horrebow and his assistants saw the satellite of Venus for the last time. However, this time he kept the observations for himself, and they only became known when they were published in 1882. Horrebow died in 1776, and was succeeded by Thomas Bugge who dismissed the satellite of Venus as a phantom.

Houzeau de Lehaie, Jean-Charles (1820–1888)

Starting as a journalist, Houzeau had an unusual career. Restless and politically active, he defended the causes of republicanism, socialism and abolition. At the same time he did astronomical work, which in 1856 made him a member of the Royal Belgian Academy. He spent the major part of the years 1857–76 abroad, mostly in America (New Orleans) and Jamaica, combining journalistic writing with astronomical observation. In 1876 he was offered the position as director of the Royal Observatory in Brussels, which he kept until 1883. Houzeau wrote on a variety of subjects, most of them unconnected with astronomy (such as sociology, technology, natural history, topography and economy). He was active in the Venus transit observations in 1882 and compiled a massive *Bibliographie générale de l'astronomie*. In 1884 he proposed that the reputed satellite of Venus was really a small planet moving a little exterior to Venus. The planet Neith (as he called it) was not received favourably and soon disappeared from the astronomical literature.

Lagrange, Louis (1711–1783)

The French-Italian Jesuit and astronomer worked at the observatory in Marseilles as assistant of Esprit Pezenas, the professor of hydraulics in Marseilles. In 1762 he was called to assist with the establishment of the new Jesuit college at Brera near

Milan, and in 1772 he was placed in charge of the Brera Observatory. He left the observatory in 1776, after having published a history of it. During this period he collaborated with Boscovich, but after 1772 the two Jesuit scientists separated and became enemies. Lagrange published his first observations in 1756, in the memoirs of the Marseille Observatory, and in May 1761 he made three observations of the satellite of Venus, but without publishing his results. They only became known through Lalande, who reported them in the *Encyclopédie*.

Lalande, Joseph Jérôme Lefrançais de (1732–1807)

Not yet twenty years old, and after having completed his studies in law, he went to Paris to observe the Moon. After a successful scientific mission to Berlin, where he met luminaries such as Voltaire, Euler and Maupertuis, in 1753, at the tender age of 21, he was elected a member of the Académie des Sciences in Paris, first as adjunct professor and from 1772 as pensionnaire. In 1768 he became professor of astronomy at the Collège Royal, where he followed his teacher Joseph-Nicolas Delisle, and in 1795 he was appointed director of the Paris Observatory. A gifted teacher and popularizer, he published many works aimed at the general public, including the eight-volume *Voyage d'un français en Italie* and the often reprinted *Astronomie des dames*. As an outspoken freethinker he collected material for the *Dictionnaire des athées* published in 1800.

Lalande's contributions to astronomy covered a wide span, both in observation, theory and organization. He served as editor of the astronomical almanac *Connaissance des temps* from 1760 until 1776 and again from 1794 until his death in 1807. He was also involved in the work on the Venus transits of 1761 and 1769, and deduced in 1770 a best value of the solar parallax. A prolific author, Lalande published numerous articles and books. His *Traité de l'astronomie* of 1764 became influential as a textbook, and his articles on astronomy in the *Encyclopédie* and *Encyclopédie méthodique* were of no less importance. He had an interest in the history of astronomy, such as shown by his *Bibliographie astronomiques* published in 1803. Another important work was the *Histoire céleste* of 1801, a star catalogue which comprised about 50,000 stars. In 1802, while director of the Paris Observatory, he established the Lalande Prize, to be awarded each year for an outstanding contribution to astronomy.

Lalande took an interest in the satellite of Venus, of which he gave an account in the *Encyclopédie*, the *Dictionnaire de physique* and elsewhere. Rather than dismissing the satellite, he maintained an uncommitted attitude and communicated the observations of Lagrange, Montaigne and others. On behalf of the Académie des Sciences, he expressed his appreciation of Baudouin's memoir of 1761 on the supposed discovery of the Venus moon. Although he did not support the actual existence of the moon, for a period he was sympathetic to the idea. By 1795 at the latest, he reached the conclusion that the satellite of Venus was not real but caused by optical illusions.

Lambert, Johann Heinrich (1728–1777)

Although born in Mulhouse in Alsace, Lambert spent most of his life in Switzerland and Germany. He became involved with science while a private teacher in Chur, Switzerland, and in 1760–61 he published in Augsburg two of his most important works, *Photometria* and the *Cosmologische Briefe*. The first was a pioneering work in photometry, the second a grand if rather speculative attempt to describe the entire stellar universe. On the recommendation of Euler, he was made a member of the Berlin Academy of Sciences in 1765. He stayed in Berlin until his death, producing important works in mathematics, philosophy and physics. Apart from *Photometria*, his main contributions to physics were included in the *Pyrometrie* of 1779, his last book. His most important mathematical work was also one of the earliest, *Die freye Perspective* published in Zurich 1759, which is a classic of descriptive geometry. In 1774 he founded the *Astronomisches Jahrbuch*, subsequently edited by J. Bode. As a true polyhistor, he covered and contributed to a large part of science and scholarship. His most important work in astronomy was a method of analyzing cometary orbits, published in 1761 as *Insigniores orbitae cometarum proprietates*. In this work he developed an earlier theory of Euler and generalized it to the case of an elliptical orbit.

In a memoir of the Berlin Academy of 1773, and also in two articles in the *Astronomisches Jahrbuch*, Lambert analyzed in detail the observations of the Venus moon. He clearly found the subject to be of value, but refrained from committing himself on the moon's existence. Having calculated its orbital elements, he predicted, wrongly, that it should be possible to see it in front of the Sun on 1 June 1777. On the basis of the moon's orbital elements he found the mass of Venus, for which he got a surprisingly large value, more than seven times the mass of the Earth.

Mairan, Jean Jacques d'Ortous de (1678–1771)

A physicist in the dying Cartesian tradition, Mairan attempted to find physical mechanisms for such phenomena as heat, light, the shape of the Earth and the aurora. He became a member of the Académie des Sciences in 1718 and served as its secretary in 1741-43. Although basically a Cartesian, he sought to construct theories which included features of Newtonian natural philosophy. Among his works was an investigation of phosphorous bodies, *Dissertation sur la cause de la lumière des phosphores et des noctiluques* from 1717, and a massive work on the northern light, *Traité physique et historiques de l'aurore boréale* from 1754. In 1741 Mairan surveyed the possible existence of the Venus moon, for which he had considerable sympathy. He suggested that the few and erratic observations were caused by an extended solar atmosphere, which hypothesis he presented in a more elaborate form in 1762. Just as he had explained comets, the aurora borealis and the zodiacal light on the basis of a subtle interplanetary medium, so he thought he could explain the observations of the satellite of Venus in the same manner.

Mayer, Andreas (1716–1782)

A German astronomer, Mayer served from 1741 as professor of mathematics, physics and astronomy at the University of Greifswald. This town in northern Germany was at the time under Swedish rule, which may explain why Mayer became a member of the Royal Swedish Academy of Sciences and at one occasion published a paper in Swedish in the society's proceedings. In 1759 he observed near Venus an object which he thought might have been a moon. Knowledge of the observation only became public after it was reported by Lambert in 1778. Mayer participated in the 1769 Venus transit project and published a paper in Latin of his observations in the *Philosophical Transactions.* He collaborated with Lampert Heinrich Röhl, who in 1773 succeeded him as director of the Greifswald Observatory. Röhl, too, was a member of the Royal Swedish Academy of Sciences.

Montaigne, Jacques Leibar (b. 1716)

Born in Narbonne, Montaigne was an accomplished amateur astronomer associated with a scientific society in Limoges, France. He made his name by discovering two comets, one in 1772 (today known as 3D/Biela) and the other in 1774. In 1780 he discovered yet another comet, subsequently verified by Heinrich Olbers. On the instigation of Baudouin, he searched in 1761 for the satellite of Venus, which he saw at four different occasions in May. Rather than communicating the results himself, Baudouin did so in two treatises to the Académie des Sciences. Because Montaigne did not publish on the matter, it is unknown how he felt about the observations and the collaboration with Baudouin.

Montbarron

A councillor in Auxerre south of Paris, he observed Venus in March 1764 and saw twice an object which he thought might have been a satellite. He was certain that it was not a star. The observations were known to Messier, who communicated them to Hell, and they also received notice in the *Encyclopédie.* Nothing further is known of Montbarron, also spelt Montbaron. He is not known to have contributed to the scientific literature.

Petit, Frédéric (1810–1865)

A student of Arago, Petit served from 1838 as director of the observatory in Toulouse, where he did research on comets, solar eclipses, meteors and bolides. A new observatory, located in Jolimont outside the city, was finished in 1850. He specialized in meteors and bolides, on which topics he published many articles in the *Comptes Rendus* and elsewhere. In 1846 and 1847, in two papers on bolides that had appeared the previous year, he concluded that it was really a new satellite of the Earth, only much smaller than the Moon. Although the suggestion was ignored by most astronomers, he continued to champion the case of a second moon until his death. It was severely criticized by Leverrier in 1851, but attracted public attention because it appeared in Jules Verne's *Autour de la lune* of 1869. The suggestion was also mentioned in Arago's posthumous *Astronomie populaire*, and Amédée Victor Guillemin, a writer of popular astronomy, supported it in his *La lune* of 1866.

Roedkiær, Peder (d. 1767)

An assistant of the astronomy professor Christian Horrebow, Roedkiær worked as an observer at the observatory of Copenhagen University. In the period June-December 1761 he observed several times what he thought was the satellite of Venus, and he saw it again in early March 1764. Probably with the support of Horrebow, he published his findings of 1764 in a communication to the proceedings of the Royal Danish Academy of Sciences (of which Horrebow, but not Roedkiær, was a member). He was convinced that what he had seen was indeed a moon of Venus. His observation of 4 March 1764 was later to attract attention. Some astronomers thought that he had unwittingly spotted the planet Uranus, while others believed he had mistaken a star for the moon of Venus; others again concluded that he had been a victim of optical illusion.

Scheuten, Abraham

A Jewish amateur astronomer, he observed on 6 June 1761 the Venus transit from his home in Crefeld, Germany. He claimed to have seen its satellite in the middle of the solar disk and assured that what he saw was not a sunspot. Scheuten's observations remained unknown until he reported them to Lambert in 1775, after which Lambert published them in his *Astronomisches Jahrbuch.*

Schorr, F.

Except from being a German amateur astronomer and member of a learned society in Danzig (now Gdansk in Poland), little is known of Schorr. In 1873 and 1875, he published two small books, one on the solar parallax problem and the other, *Der Venusmond*, on the satellite of Venus. He provided valuable historical information and argued that the accumulated evidence was in favour of the existence of a satellite. Although no astronomer accepted his claim, the book was fairly well received and helped to revive interest in a possible Venus moon, or at least to explain the earlier observations. Apart from his two books, Schorr seems not to have contributed to the astronomical literature.

Schröter, Johann Hieronymus (1745–1816)

An observational astronomer, Schröter's strength lay in his enthusiasm for observation and his excellent instruments. As chief magistrate of Lilienthal, near Bremen, he spent his free time with observations. His mirror telescope completed in 1793 was the largest in Germany. Tragically, when French troops invaded Lilienthal in 1813, his observatory was burned down. Schröter is best known for his observations of the Moon and Venus. He published a book in two volumes on lunar topography, the *Selenotopographische Fragmente* (1791–97), and collected his studies of Venus in the *Aphroditographische Fragmente* (1796). In 1790 he established the presence of an atmosphere around Venus, and he also claimed to have seen very high mountains on the planet. Although he looked for the satellite of Venus, he never saw it and also he saw no secondary images of the kind that Hell had suggested. Together with his assistant Karl Harding, in 1797 he succeeded in observing the two moons of Uranus discovered by William Herschel.

Short, James (1710–1768)

Born in Edinburgh, he entered the university in 1726 and became the protégé of Colin Maclaurin, the professor of mathematics. He started making mirrors for telescopes at an early age. In 1737 he was elected a fellow of the Royal Society, and the following year he settled in London, where he established his reputation as a first-rate optician and astronomer. He specialized in reflecting telescopes, of which he made a total of 1,370. In 1757 he became a foreign member of the Royal Swedish Academy of Sciences. He was one of the first British natural philosophers to endorse the vibration theory of light and to incorporate John Dollond's achromatic lens into a telescope. Short was actively involved in the determination of the solar parallax based on the Venus transit in 1761, on which subject he published two papers. He was appointed to a special committee set up by the Royal Society to study the 1769 transit, but died before he could take part in the plans. In 1740 he observed an object near to Venus which he described as a "small star." He may or may not have thought it was a satellite, but later in life he seems to have denied the existence of a Venus moon. In any case, his report in the *Philosophical Transactions* was well known, also on the Continent, and caused other astronomers to consider the possibility of a satellite of Venus.

Stroobant, Paul Henri (1868–1936)

After having worked as a voluntary assistant at the Royal Observatory at Uccle near Brussels, he presented in 1887 (at the age of nineteen) a systematic analysis of the satellite of Venus to the Royal Belgian Academy. According to Stroobant, almost all of the sightings had been of faint stars, a claim which he substantiated by comparing the positions of the alleged satellite with the *Bonner Durchmusterung* star catalogue. He concluded that there was no basis at all for believing in a Venus moon, which conclusion was generally accepted. Most astronomers thought that he had solved the mystery of the satellite of Venus.

In 1889 Stroobant earned a doctorate in physics and mathematics from the University of Brussels. After studies in Paris, he returned to Brussels, where he became professor of astronomy in 1896 and director of the Royal Observatory in 1925. During his distinguished career, he worked in a variety of fields, including the satellites of Saturn, the personal equation and the dynamics and distribution of stars in the Milky Way. He was active in the International Astronomical Union, where he served as president for its bibliographical committee. Among his honours was the Lalande Prize, which he received from the Paris Académie des Sciences in 1921. In 1950 a Stroobant Prize was created by the Royal Belgian Academy of Sciences, to be awarded to an outstanding Belgian or French astronomer.

Bibliography

Note: A most valuable primary source is the *Histoire et Mémoires de l'Académie Royale des Science* published by the Paris Academy in 93 volumes between 1699 and 1790. This series consists of two individually paginated parts appearing in the same volumes. They are here denoted *Histoire* and *Mémoires*, respectively (Halleux et al. 2001). Journal articles are listed according to the year of publication, which in many cases is later than the year corresponding to the volume number. Another important source, the *Astronomische Nachrichten* established by Heinrich Christian Schumacher in 1823, was organized in columns rather than pages, each page comprising two columns.

Aikin, Arthur (1804). [Review of Thornton 1803]. *Annual Review and History of Literature* (for 1803) **2**, 873–882.

Alemi, Alex and David Stevenson (2006). "Why Venus has no moon," *Bulletin of the American Astronomical Society* **38**, 491.

Alexander, Arthur F. O'Donel (1965). *The Planet Uranus: A History of Observation, Theory and Discovery.* London: Faber and Faber.

Anger, Karl T. (1862). *Populäre Vorträge über Astronomie.* Ed. by Gustav Baddach. Danzig: Kasemann.

Anon. (1876a). "Short's observation of a supposed satellite of Venus," *Nature* **14**, 231–232.

Anon. (1876b). "The planet Vulcan," *Littel's Living Age* **131** (issue 1690), 318–320.

Anon. (1884). "The problematical satellite of Venus," *The Observatory* **7**, 222–226.

Anon. (1887a). "The 'satellite' of Venus," *Nature* **36**, 543.

Anon. (1887b). "The supposed satellite of Venus," *The Observatory* **10**, 363–364.

Arago, D. François J. (1854–60). *Astronomie populaire*, 4 vols. Paris: Gide et Baudry.

Artzt, Friederich (1813). "Iagttagelse over Venus og dennes Drabant," *Nyeste Skilderie af Kjøbenhavn* **29**, 452–455.

Ashbrook, Joseph (1954). "The satellite of Venus," *Sky and Telescope* **13**, 333.

— (1955). "Other satellites of the Earth," *Sky and Telescope* **14**, 334.

— (1984). *The Astronomical Scrapbook: Skywatchers, Pioneers, and Seekers in Astronomy.* Cambridge, Mass.: Sky Publishing Corporation.

Ashmore, Malcolm (1993). "The theatre of the blind: Starring a Promethean prankster, a phoney phenomenon, a prism, a pocket, and a piece of wood," *Social Studies of Science* **23**, 67–106.

Astore, William J. (2001). *Observing God: Thomas Dick, Evangelicalism, and Popular Science in Victorian Britain and America.* Aldershot: Ashgate.

Austin, R. H. (1967). "Uranus observed," *British Journal for the History of Science* **3**, 275–284.

Bailly, Jean-Sylvain (1779). *Histoire de l'astronomie moderne depuis la fondation de l'école d'Alexandrie*, 2 vols. Paris: Freres de Bure.

Bakich, Michael E. (2000). *The Cambridge Planetary Handbook.* Cambridge: Cambridge University Press.

Barnard, Edward E. (1892). "Discovery and observations of a fifth satellite to Jupiter," *Astronomical Journal* **12**, 81–85.

— (1895). "On a photographic search for a satellite to the Moon," *Astrophysical Journal* **2**, 347–349.

— (1906). "An unexplained observation," *Astronomische Nachrichten* **173**, no. 4106, columns 25–26.

Baudouin, Armand H. (1761a). *Mémoire sur la découverte du satellite de Vénus, et sur les nouvelles observations qui viennent d'être faites à cette sujet.* Paris: Desaint & Saillant.

— (1761b). *Remarques sur une quatriéme observation du satellite de Venus faite à Limoges le 11 mai 1761.* Paris: Desaint & Saillant.

— (1761c). *Abhandlung von der Entdeckung eines Trabanten der Venus.* Berlin: Verlag des Buchladens der Realschule.

— (1768a). "Mémoire sur la position de l'orbite de Vénus dans son passage sur le soleil du 6 Juin dernier, en y employant les observations de Gottingen," *Mémoires de mathématiques et de physique* (Recueils des savants Étrangers) **5**, 270–275.

— (1768b). "Observations des troisiéme et quatriéme satellites de Jupiter," *Mémoires de mathématiques et de physique* (Recueils des savants Étrangers) **5**, 124–127.

Baum, Richard (1973). *The Planets: Some Myths and Realities.* Newton Abbot: David and Charles.

— (1999). "Star-spots on Venus," *Journal of the British Astronomical Association* **109**, 115–116.

— (2000). "The enigmatic ashen Light of Venus: An overview," *Journal of the British Astronomical Association* **110**, 325–329.

Baum, Richard and William Sheehan (1997). *In Search of Planet Vulcan. The Ghost in Newton's Clockwork Universe.* Cambridge, Mass.: Basic Books.

Bedini, Silvio A. (1961). "The optical workshop equipment of Guiseppe Campani," *Journal of the History of Medicine and Allied Sciences* **16**, 18–38.

Bernoulli, Jean (1771). *Recueil pour les astronomes*, vol. 2. Berlin: G. J. Decker.

— (1802). "Muthmassung über den Uranus. Aus einem Schreiben des Direktors Bernoulli in Berlin," (Zach's) *Monatliche Correspondenz zur Beförderung der Erd- und Himmelkunde* **5**, 343–344.

Bertrand, Joseph (1875). "Le satellite de Vénus," *Journal des Savants*, 456–461.

— (1882). "Le satellite de Vénus," *L'Astronomie* **1**, 201–206.

Bianchini, Francesco (1996). *Observations Concerning the Planet Venus.* Trans. by Sally Beaumont and Peter Gay. Berlin: Springer-Verlag.

Bigourdan, G. (1904). "Sur les observations de comètes de Messier," *Bulletin Astronomique* **21**, 121–128.

Blacklock, Arthur W. (1868). "The satellite of Venus," *Astronomical Register* **6**, 196–197.

Bliss, Nathanael (1762). "A second letter ... concerning the transit of Venus over the Sun, on the 6th of June 1761," *Philosophical Transactions* (for 1761) **52**, 232–250.

Blumenberg, Hans (1975). *Die Genesis der kopernikanischen Welt.* Frankfurt am Main: Suhrkamp.

Bode, Johann E. (1807). *Allgemeine Betrachtungen über das Weltgebäude.* Berlin: Kechtschen Buchhandlung.

— (1816). *Betrachtung der Gestirne und des Weltgebäudes.* Berlin: Nicolaischen Buchhandlung.

Bonnet, Charles (1764). *Contemplation de la nature.* Amsterdam: M. M. Ray.

— (1766). *Betrachtung über die Natur, vom Herrn Karl Bonnet.* Trans. by J. D. Titius. Leipzig: J. F. Junius.

Bopp, K. (1915). *Johann Heinrich Lamberts Monatsbuch mit den zugehörigen Kommentaren.* Munich: Verlag der Königlich Bayerischen Akademie der Wissenschaften.

Boscovich, Roger J. (1761). "De proximo Veneris sub Sole transitu," *Philosophical Transactions* (for 1760) **51**, part 2, 865–888.

— (1767). *Dissertationes quinque ad dioptricam pertinentes.* Vienna: Joannis Thomæ de Trattnern.

Brasch, F. E. (1916). "John Winthrop (1714–1779), America's first astronomer, and the science of his period," *Publications of the Astronomical Society of the Pacific* **28**, 153–170.

Bravais, Auguste (1847). "Mémoir sur les halos et les phénomènes optique qui les accompagnent," *Journal de l'École Royale Polytechnique* **18**, cahier 31, 1–280.

Brewster, David (1854). *More Worlds than One: The Creed of the Philosopher and the Hope of the Christian.* London: John Murray.

Briggs, J. Morton (1967). "Aurora and Enlightenment: Eighteenth-century explanations of the aurora borealis," *Isis* **58**, 491–503.

Brooke, John H. (1977). "Natural theology and the plurality of worlds: Observations on the Brewster–Whewell debate," *Annals of Science* **34**, 221–286.

Brush, Stephen G. (1996). *Nebulous Earth. The Origin of the Solar System and the Core of the Earth from Laplace to Jeffreys.* Cambridge: Cambridge University Press.

Brückner, Johannes A. (1828). *Gemeinverständliche Astronomie*, 2 vols. Leipzig: Reclam.

Bugge, Thomas (1783). "Om den nye planet," *Skrifter, Kgl. danske Videnskabernes Selskab*, Part 2 (for 1781), 215–225.

Burbidge, Gaston (1958). *Rosicrucian Digest 1958.* San Jose, CA: Kessinger Publications.

Burke, John G. (1986). *Cosmic Debris: Meteorites in History.* Berkeley: University of California Press.

Burns, Joseph A. (1973). "Where are the satellites of the inner planets?" *Nature, Physical Science* **242**, 23–25.

Bush, Clive (1974). "Erasmus Darwin, Robert John Thorncroft, and Linnaeus' sexual system," *Eighteenth-Century Studies* **7**, 295–320.

Carrington, Richard C. (1860). "On some previous observations of supposed planetary bodies in transit over the Sun," *Monthly Notices of the Royal Astronomical Society* **20**, 192–194.

Cassini, Jacques (1735). "De la révolution de Vénus autour de son axe," *Mémoires de l'Académie Royale des Sciences* (for 1732) **34**, 197–214.

— (1740). *Elémens astronomie.* Paris: Imprimierie Royale.

Cassini, Jean Dominique (1730). "Découverte de la lumiere celeste qui paroist dans le Zodiaque," *Mémoires de l'Académie Royale des Sciences depuis 1666 jusqu'a 1699* **8**, 180–278.

Cassini, Jean Dominique and Giacomo Filipo Maraldi (1707). "Taches dans le soleil observées le 29. Mars 1701," *Mémoires de l'Académie Royale des Sciences* (for 1701) **3**, 101–103.

Cassini de Thury, César-François (1763). "Observation du passage de Vénus sur le soleil faites à Vienne en Autriche," *Histoire de l'Académie Royale des Sciences* (for 1761) **63**, 409–412.

Chapin, Seymour L. (1978). "Lalande and the longitude: A little known London voyage of 1763," *Notes and Records of the Royal Society of London* **32**, 165–180.

Chappe d'Auteroche, Jean-Baptiste (1772). *Voyage en Californie pour l'observation du passage de Vénus sur le disque du Soleil, le 3 Juin 1769.* Ed. by César-François Cassini de Thury. Paris: Charles-Antoine Jombert.

Chapple, William (1761a). "Observation of the transit at Exeter," *Gentleman's Magazine* **31** (June), 252.

— (1761b). "An observation of the transit of Venus rectify'd," *Gentleman's Magazine* **31** (August), 357–359.

Clerke, Agnes M. (1903). *Problems in Astrophysics.* London: Adams & Charles Black.

Cohen, I. Bernard (1978). "Perfect numbers in the Copernican system: Rheticus and Huygens," pp. 419–425 in Erna Hilfstein, ed., *Science and History. Studies in Honor of Edward Rosen* (*Studia Copernicana* **16**). Warsaw: Polish Academy of Sciences Press.

— (1990). "G. D. Cassini and the number of the planets: An example of seventeenth-century astro-numerological patronage," pp. 199–206 in Trevor H. Levere and William R. Shea, eds., *Nature, Experiment, and the Sciences: Essays on Galileo and the History of Science in Honour of Stillman Drake.* Dordrecht: Kluwer Academic.

Constance, Arthur (1956). *The Inexplicable Sky.* New York: Citadel Press.

Copeland, Ralph (1892). "On a pretended early discovery of a satellite of Mars," *Monthly Notices of the Royal Astronomical Society* **52**, 493–494.

Corliss, William R. (1979). *The Mysterious Universe: A Handbook of Astronomical Anomalies.* Glen Arm, Maryland: Sourcebook Project.

Crowe, Michael J. (1999). *The Extraterrestrial Life Debate, 1750–1900.* Mineola, New York: Dover Publications.

d'Alembert, Jean le Rond (1822). *Oeuvres de d'Alembert*, vol. 5, part 1. Paris: A. Belin.

Dal Prete, Ivano (2005). "Francesco Bianchini e il pianeta Venere," *Nuncius* **20**, 95–152.

Darnton, Robert (1979). *The Business of Enlightenment: A Publishing History of the Encyclopédie 1775–1800.* Cambridge, Mass.: Harvard/Belknap.

D'Arrest, Heinrich L. (1865). "Schreiben des Herrn Prof. d'Arrest," *Astronomische Nachrichten* **64**, no. 1517, columns 73–76.

Daston, Lorainne and Peter Galison (2007). *Objectivity.* New York: Zone Books.

Débarbat, Suzanne and Curtis Wilson (1989). "The Galilean satellites of Jupiter from Galileo to Cassini, Rømer and Bradley," pp. 144–157 in René Taton and Curtis Wilson, eds., *Planetary Astronomy from the Renaissance to the Rise of Astrophysics. Part A: Tycho Brahe to Newton.* Cambridge: Cambridge University Press.

Delambre, Jean-Baptiste J. (1821). *Histoire de l'astronomie moderne*, vol. 2. Paris: V. Courcier.

Denning, Wm. Frederick (1882). "Observations of Venus in the spring of 1881," *Monthly Notices of the Royal Astronomical Society* **42**, 109–112.

Derham, William (1721). *Astro-Theology or, a Demonstration of the Being and Attributes of God from a Survey of the Heavens.* London: W. and J. Immys.

Dick, Steven J. (1988). "Discovering the moons of Mars," *Sky and Telescope* **76**, 242–243.

Dick, Thomas (1838). *Celestial Scenery; or, the Wonders of the Planetary System Displayed; Illustrating the Perfections of Deity and a Plurality of Worlds.* New York: Harper & Brothers.

Dionis du Séjour, Achille-Pierre (1789). *Traité analytique des mouvemens apparens des corps célestes*, vol. 2. Paris: La Veuve Valade.

Dunn, Samuel (1763). "Some observations of the planet Venus, on the disk of the sun, June 6th, 1761," *Philosophical Transactions* (for 1761), **52**, 184–195.

Eames, John (1735). "An account of a book entitled, Traité physique et historique de l'aurore boreale, par Mr de Mairan," *Philosophical Transactions* (for 1733–34), **38**, 243–257.

Eisenstaedt, Jean (1991). "De l'influence de la gravitation sur la propagation de la lumière en théorie newtonienne: L'archéologie des trous noirs," *Archive for History of Exact Sciences* **42**, 315–386.

[Encyclopédie I] (1751–72). *Encyclopédie ou dictionnaire raisonné des sciences, des arts et des métiers*, 17 vols. Ed. by Denis Diderot and Jean d'Alembert. Paris: Briasson et al.

[Encyclopédie II] (1778–82). *Encyclopédie ou dictionnaire raisonné des sciences, des arts et des métiers*, 36 vols. Ed. by Jean-Antoine de Laserre. Berne and Lausanne: Sociétés Typographiques.

Ende, Ferdinand A. (1811). "Auszug aus einem Schreiben des Freyherrn von Ende," (Zach's) *Monatliche Correspondenz zur Beförderung der Erd- und Himmelkunde* **24**, 389–394.

Erxleben, Johann C. P. (1787). *Anfangsgründe der Naturlehre.* Ed. by Georg C. Lichtenberg. Göttingen: Johann C. Dieterich.

Euler, Leonhard (1862). *Leonhardi Euleri, Opera Postuma, Mathematica et Physica.* Ed. by P. H. Fuss and N. Fuss. St. Petersburg: Eggers et Socios.

Farley, John (1811). "On the nature of those meteors called shooting stars," (Nicholson's) *Journal of Natural Philosophy, Chemistry, and the Arts* **29**, 285–286.

Faure, Georges le and Henry de Graffigny (1888). *Aventures extraordinaires d'un savant Russe.* Vol. 2: *Le soleil et les petites planètes.* Paris: Édinger.

Ferguson, James (1778). *Astronomy Explained upon Sir Isaac Newton's Principles.* London: W. Strathan.

Ferner, Bengt (1763). "An account of the observations of the same transit made in and near Paris," *Philosophical Transactions* (for 1761), **52**, 221–225.

Fischer, Ernst G. (1787). "Ueber das Monden-System des Uranus," *Astronomisches Jahrbuch oder Ephemeriden* (for 1790), 213–219.

Flammarion, (1880). *Astronomie populaire.* Paris: Marpon et Flammarion.

— (1884). *Les terres du ciel: Voyages astronomique sur les autres mondes.* Paris: C. Marpon et E. Flammarion.

Fleischer, Esaias (1786). *Forsøg til en Natur-Historie*, vol. 1. Copenhagen: Gyldendal.

Fontana, Francesco (1646). *Novae coelestium terrestriumque rerum observationes.* Naples: Gaffaro.

— (2001). *New Observations of Heavenly and Earthly Objects, with the Aid of Optical Instruments devised by Him and Brought to the Highest state of Perfection.* Trans. and ed. by Peter Fay and Sally Beaumont. Privately printed.

Fontenelle, Bernard le Bouvier de (1740). *Éloges des académiciens. Avec l'histoire de l'Académie Royale des Sciences*, 2 vols. The Hague: Isaac vander Kloot.

— (1955). *Entretiens sur la pluralité des mondes.* Ed. by Robert Shackelton. Oxford: Clarendon Press.

Forbes, Eric G. (1982). "The pre-discovery of Uranus," pp. 67–79 in Gary E. Hunt, ed., *Uranus and the Outer Planets.* Cambridge: Cambridge University Press.

Fort, Charles H. (1999). *The Book of the Damned.* Amherst, New York: Prometheus Books.

Fouchy, Grandjean de (1764). "Sur le satellite vu ou présumé autour de Vénus," *Histoire de l'Académie Royale des Sciences* (for 1762) **64**, 116–120.

— (1774). "Éloge de M. de Mairan," *Histoire de l'Académie Royale des Sciences* (for 1771) **74**, 89–104.

Frederick II (1805). *Oeuvres posthumes de Fréderic II, roi de prusse. Correspondance avec M. d'Alembert*, vol. 1. Potsdam: Des Associés.

Galiani, Ferdinand (1818). *Correspondance inédite de l'abbé Ferdinand Galiani, conseiller du roi de Naples.* Paris: Treuttel and Wurtz.

Galilei, Galileo (1929–39). *Le Opere di Galileo Galilei*, 20 vols. Ed. by Antonio Favaro. Florence: S. Barbéra.

— (1989). *Sidereus Nuncius, or the Sidereal Messenger.* Trans. and notes by Albert van Helden. Chicago: University of Chicago Press.

Gardner, Martin (1957). *Fads and Fallacies in the Name of Science.* New York: Dover Publications.

Garisto, Robert (1991). "An error in Isaac Newton's determination of planetary properties," *American Journal of Physics* **59**, 42–48.

Gassendi, Pierre (1997). *Institution astronomique.* Trans. by Jean Peyroux. Paris: Librairie A. Blanchard.

Gehler, Johann S. T. (1798). *Physikalisches Wörterbuch*, vol. 3. Leipzig: Schwickert.

Gingerich, Owen (1970). "The satellites of Mars: Prediction and discovery," *Journal for the History of Astronomy* **1**, 109–115.

— (1978). "The discovery of the satellites of Mars," *Vistas in Astronomy* **22**, 127–132.

Goldstein, Bernie R. (1969). "Some medieval reports of Venus and Mercury transits," *Centaurus* **14**, 49–59.

Gore, J. Ellard (1909). *Astronomical Curiosities. Facts or Fallacies.* London: Chatto & Windus.

Gould, S. H. (1945). "Gulliver and the moons of Mars," *Journal of the History of Ideas* **6**, 91–101.

Greene, John C. (1954). "Some aspects of American astronomy 1750–1815," *Isis* **45**, 339–358.

Gregory, David (1702). *Physicæ & geometricæ elementa.* Oxford: Sheldonian Theatre.

— (1736). *Elements of Physical and Geometrical Astronomy*, 2 vols. London: D. Midwinter. New York: Johnson Reprint Corporation 1972.

Grosser, Morton (1979). *The Discovery of Neptune.* New York: Dover Publications.

Guericke, Otto von (1968). *Otto von Guerickes neue (sogenannte) Magdeburger Versuche über den leeren Raum.* Ed. and trans. by Hans Schimank. Düsseldorf: VDI-Verlag.

— (1994). *The New (So-Called) Magdeburg Experiments of Otto von Guericke.* Ed. and trans. by Margaret G. F. Ames. Dordrecht: Kluwer Academic.

Guillemin, Amédée V. (1866). *La lune.* Paris: Hachette.

Haase, C. (1863–69). "Einige Zusammenstellungen als Beitrag zur der Frage, ob ausser Mercur und Venus in dem Raume zwischen Sonne und Erde noch andere planetenartige Körper vorhanden sind," [Peter's] *Zeitschrift für populäre Mittheilungen aus dem Gebiete der Astronomie und verwandter Wissenschaften* **2** (1863), 165–256; **3** (1869), 1–56.

Hahn, Friedrich von (1801). "Beobachtung eines kleinen beweglichen Sterns, sehr nahe bey dem veränderlichen Stern Mira am Halse des Wallfisches," *Astronomisches Jahrbuch oder Ephemeriden* (for 1804), 195–197.

Hall, Asaph (1878). "The discovery of the satellites of Mars," *Monthly Notices of the Royal Astronomical Society* **38**, 205–209.

Halleux, Robert et al. (2001). *Les publications de l'Académie Royale des Sciences de Paris (1666–1793)*, 2 vols. Turnhout: Brepols.

Hankins, Thomas L. (1993). "The ocular harpsichord of Louis-Bertrand Castel; or, the instrument that wasn't," *Osiris* **9**, 141–156.

Heilbron, John L. (1999). *The Sun in the Church: Cathedrals as Solar Observatories.* Cambridge, Mass.: Harvard University Press.

— (2005). "Bianchini as an astronomer," pp. 57–82 in Valentin Kockel and Brigitte Sölch, eds., *Francesco Bianchini (1662–1729) und die europäische gelehrte Welt um 1700.* Berlin: Akademie Verlag.

Hell, Maximilian (1765). *De sattelite Veneris.* Vienna: Joannis Thomæ de Trattnern.

— (1766). *Ephemerides astronomicæ meridianum Vindobonensem.* Vienna: Joannis Thomæ de Trattnern.

— (1792). *Beyträge zur praktischen Astronomie*, vol. 2. Trans. by Longinus A. Jungnitz. Breslau: Korn.

Helmuth, Johann H. (1794). *Anleitung zur Kenntniss des grossen Weltbaues für Frauenzimmer in freundschaftlichen Briefen.* Braunschweig: Schulbuchhandlung.

Hennesey, R. A. S. (1999). *Worlds Without End. The Historic Search for Extraterrestrial Life.* Stroud: Tempus Publishing.

Herder, Johann G. (1787). *Gott. Einige Gespräche über Spinoza's System.* Gotha: K. W. Ettinger.

Herschel, John (1833–36). "On the satellites of Uranus," *Monthly Notices of the Royal Astronomical Society* **3**, 35–36.

Herschel, William (1912). *The Scientific Papers of Sir William Herschel*, 2 vols. London: The Royal Society.

Hetherington, Norriss S. (1988). *Science and Objectivity: Episodes in the History of Astronomy.* Ames: Iowa State University Press.

Hevelius, Johannes (1674). "A certain phenomenon, seen by Monsieur Havelius, Feb. 5, 1674," *Philosophical Transactions* **9**, no. 101, 26–27.

— (1967). *Selenographia sive Lunae descriptio.* New York: Johnson Reprint Corporation.

Hill, Elizabeth (1961). "Biographical essay," pp. 17–101 in Lancelot L. Whyte, ed., *Roger Joseph Boscovich: Studies in his Life and Work on the 250th Anniversary of his Birth.* London: Allen & Unwin.

Hind, John R. (1852). *The Solar System: A Descriptive Treatise upon the Sun, Moon, and Planets.* New York: G. P. Putnam.

Hockey, Thomas, ed. (2007). *Biographical Encyclopedia of Astronomers*, 2 vols. New York: Springer.

Horrebow, Christian (1761a). *Dissertatio de semita, quam in Sole descripsit Venus per eundem transeunde die 6 Junii Ao. 1761.* Copenhagen: N. C. Höpffneri.

— (1761b). *Dissertatio de transitu Veneris per discum Solis.* Copenhagen: N. C. Höpffneri.

— (1765a). "Tidens bestemmelse i henseende til de observationer, som skeede i Solen og Venere den 6. Juni 1761," *Skrifter, Kgl. danske Videnskabernes Selskab*, Part 9 (for 1761–64), 373–388.

— (1765b). "Reflexioner anlangende Veneris drabant," *Skrifter, Kgl. danske Videnskabernes Selskab,* Part 9 (for 1761–64), 396–399.

— (1765c). "Videre fortsættelse af observationerne gjorte paa Veneris drabant," *Skrifter, Kgl. danske Videnskabernes Selskab*, Part 9 (for 1761–64), 400–403.

Houzeau, Jean-Charles (1878). "Sur certains phénomènes énigmatiques de l'astronomie," *Bulletin de l'Académie Royale (de Belgique) des Sciences, des Lettres et des Belles Arts* **46**, 951–966.

— (1884). "Le satellite problématique de Vénus," *L'Astronomie: Revue d'astronomie populaire* **3**, 283–289.

Houzeau, Jean-Charles and Albert Lancaster, eds. (1964). *Bibliographie générale de l'astronomie jusqu'en 1880*, 2 vols. Ed. by D. W, Dewhirst. London: Holland Press. Revised reprint of original: Brussels: X. Havermans, 1882–89.

Hoyt, Douglas V. and Kenneth H. Schatten (1995). "A new interpretation of Christian Horrebow's sunspot observations from 1761 to 1777," *Solar Physics* **160**, 387–392.

Hoyt, William G. (1976). "W. H. Pickering's planetary predictions and the discovery of Pluto," *Isis* **67**, 551–564.

Humboldt, Alexander von (1845–58). *Kosmos. Entwurf einer physischen Weltbeschreibung*, 4 vols. Stuttgart: Cotta'scher Verlag.

Hunt, Gary E. and Patrick Moore (1982). *The Planet Venus*. London: Faber and Faber.

Hussey, Thomas J. (1834). "On the rotation of Venus," *Astronomische Nachrichten* **11**, no. 248, columns 121–146.

Hutton, Charles (1795). *A Mathematical and Philosophical Dictionary*, 2 vols. London: J. Johnson.

Huygens, Christiaan (1722). *The Celestial Worlds Discover'd: or, Conjectures Concerning the Inhabitants, Plants and Productions of the Worlds in the Planets*. London: James Knapton.

— (1888). *Œuvres complètes de Christiaan Huygens*, vol. 1. The Hague: Martinus Nijhoff.

— (1925). *Œuvres complètes de Christiaan Huygens*, vol. 15. The Hague: Martinus Nijhoff.

— (1944). "Kosmoteoros, sive de terries cœlestibus, earumque ornatu conjecturæ," pp. 676–821 in *Œuvres complètes de Christiaan Huygens*, vol. 21. The Hague: Martinus Nijhoff.

Israel, Werner (1987). "Dark stars: The evolution of an idea," pp. 199–276 in S. Hawking and W. Israel, eds., *Three Hundred Years of Gravitation*. Cambridge: Cambridge University Press.

Jaki, Stanley L. (1972a). "The early history of the Titius-Bode law," *American Journal of Physics* **40**, 1014–1023.

— (1972b). "The original formulations of the Titius–Bode law," *Journal for the History of Astronomy* **3**, 136–138.

John, Klaus (2004). *August Heinrich Christian Gelpke – ein Astronom am Collegium Carolinum zu Braunschweig.* Doctoral dissertation, Braunschweig Technical University.

Johnson, J. H. (1930–31). "The discovery of the first four satellites of Jupiter," *Journal of the British Astronomical Association* **49**, 164–171.

Joy, Alfred H. (1959). "Mira Ceti," *Astronomical Society of the Pacific Leaflet* **8**, 57–64.

Kaiser, Frederik (1867). *Stjernehimlen.* Trans. by Mathilde Ørsted. Copenhagen: T. Lind.

Kästner, Abraham G. (1841). *Gesammelte poetische und prosaische schönwissenschaftliche Werke*, Part 1. Berlin: T. C. F. Enslin.

Kant, Immanuel (1981). *Universal Natural History and Theory of the Heavens.* Trans. and notes by Stanley L. Jaki. Edinburgh: Scottish University Press.

Keil, Inge (2000). *Augustanus Opticus: Johann Wiesel (1583–1662) und 200 Jahre optisches Handwek in Augsburg.* Berlin: Akademie Verlag.

Kepler, Johannes (1965). *Kepler's Conversation with Galileo's Sideral Messenger.* Trans. and notes by Edward Rosen. New York: Johnson Reprint Corporation.

Kindermann, Eberhard C. (1746). *Astronomische Beschreibung und Nachricht von dem Cometen 1746.* Dresden: G. C. Hilschern.

— (2006). *Die geschwinde Reise auf dem Luft-Schiff nach der obern Welt.* Zurich: De Jong Publications.

Kircher, Athanasius (1660). *Iter extaticum coeleste.* Nuremberg: J. A. Endteorum.

Klein, Hermann J. (1879). *Anleitung zur Durchmusterung des Himmels. Astronomische Objecte für gewöhnliche Teleskope.* Braunschweig: Vieweg und Sohn.

— (1887). "Das Rätsel des Venusmondes gelöst," *Sirius: Zeitschrift für populäre Astronomie* **20**, 243–249.

Kleinbaum, Abby R. (1973). *Jean Jacques Dortous de Mairan (1678–1771): A Study of an Enlightenment Scientist* (Ann Arbor: University Microfilms 73–16,215).

Koch, Julius A. (1802). "Aus einem Schreiben des Hrn. Doctor und Astronom Koch in Danzig," *Astronomisches Jahrbuch oder Ephemeriden* (for 1805), 233–236.

Koyré, Alexandre (1961). *La révolution astronomique: Copernic, Kepler, Borelli.* Paris: Hermann.

Krafft, Fritz (2002). "Was die Welt zusammenhält: Das astronomisch-physikalische Weltbild Otto von Guerickes," pp. 90–107 in Matthias Puhle, ed., *Die Welt im leeren Raum: Otto von Guericke 1602–1686.* Munich: Deutscher Kunstverlag.

Kragh, Helge (1998). "Social constructivism, the gospel of science, and the teaching of physics," pp. 125–137 in Michael R. Matthews, ed., *Constructivism in Science Education.* Dordrecht: Kluwer Academic.

— (2008). "The second moon of the Earth," *Journal for the History of Astronomy* **39** (forthcoming).

Krisch, August, ed. (1905). *Astronomisches Lexikon.* Vienna: A. Hartleben.

Kritzinger, H. H. (1926). "Das Rätsel des zweiten Erdmondes," *Sirius* **59**, 184–186.

Lacaille, Nicholas-Louis de (1763). "Observation du passage de Vénus sur le disque du soleil," *Histoire de l'Académie Royale des Sciences* (for 1761), 78–81.

Lalande, J. Jérôme (1763a). "Remarques sur les observations du passage de Vénus faites à Copenhague et à Drontheim en Norvège," *Mémoires de l'Academie Royale des Sciences* (for 1761) **63**, 113–114.

— (1763b). "Observation du passages de Vénus sur le disque du Soleil," *Mémoires de l'Academie Royale des Sciences* **63** (for 1761), 81–87.

— (1773). "Explication du prolongement obscur du disque de Vénus, qu'on aperçoit dans ses passages sur le Soleil," *Mémoires de l'Académie Royale des Sciences* (for 1770) **70**, 406–412.

— (1781). "Satellites de Vénus," pp. 250–260 in [*Encyclopédie II*], vol. 35.

— (1792). *Astronomie*, 3 vols. Paris: Didot l'Ainé.

— (1795). *Abrégé d'astronomie.* Paris: Firmin Didot.

— (1803). *Bibliographie astronomique, avec l'histoire de l'astronomie depuis 1781 jusqu'a 1802.* Paris: Imprimerie de la Republique.

Lambert, Johann H. (1773). "Essai d'une théorie du satellite de Vénus," *Nouveaux Mémoires de l'Academie Royale des Sciences et Belles-Lettres Année 1773* (Classe de Philosophie Expérimentale), 222–250.

— (1775). "Vom Trabanten der Venus," *Astronomisches Jahrbuch oder Ephemeriden* (for 1777), 186–191.

— (1776). "Vom Trabanten der Venus," *Astronomisches Jahrbuch oder Ephemeriden* (for 1778), 178–188.

— (1976). *Cosmological Letters on the Arrangement of the World Edifice.* Trans. and notes by Stanley L. Jaki. Edinburgh: Scottish University Press.

Lamont, Johann (1838). "Value of the mass of Uranus," *Monthly Notices of the Royal Astronomical Observatory* **4**, 122–124.

Laplace, Pierre Simon (1813). *Exposition du système du monde,* 4th edn. Paris: V. Courcier.

Lardner, Dionysius (1860). *Handbook of Astronomy,* 2nd edn. Ed. by Edwin Dunkin. London: Walton & Maberly.

Lassell, William (1852). "On the interior satellites of Uranus," *Monthly Notices of the Royal Astronomical Society* **12**, 15–17.

Lecomte, Stéphane (1990). "Du satellite de Vénus," *Observations et Travaux* **22**, 55–56.

Lemonnier, Pierre C. (1746). *Institutions astronomique, ou leçons élémentaires d'astronomie.* Paris: H. Guérin & J. Guérin.

Leverington, David (2003). *Babylon to Voyager and Beyond: A History of Planetary Astronomy.* Cambridge: Cambridge University Press.

Leverrier, Urbain (1851). "Remarques à l'occasion de la dernière communication de M. Petit sur les bolides," *Comptes Rendus de Séances de l'Académie des Sciences* **32**, 561–566.

Ley, Willy (1964). *Watchers of the Sky: An Informal History of Astronomy from Babylon to the Space Age.* New York: Viking Press.

Littrow, Joseph J. (1886). *Wunder des Himmels, oder gemeinfassliche Darstellung des Weltsystemes,* 7th edn. Ed. and augmented by Edmund Weiss. Berlin: G. Hempel.

Lovejoy, Arthur O. (1964). *The Great Chain of Being. A Study of the History of an Idea.* Cambridge, Mass.: Harvard University Press.

Lubbock, Constance A. (1933). *The Herschel Chronicle.* Cambridge: Cambridge University Press.

Luplau Janssen, Carl (1928). *Er Stjernerne Beboet?* Copenhagen: Gyldendal.

Lynn, W. T. (1884). "Montaigne of Limoges," *The Observatory* **7**, 231–232.

— (1887a). "Montaigne's alleged observation of a satellite of Venus in 1761," *The Observatory* **10**, 73–74.

— (1887b). "The comet discovered by Montaigne and Olbers in 1780," *The Observatory* **10**, 355–356.

Macpherson, Hector (1906). *A Century's Progress in Astronomy.* Edinburgh: William Blackwood and Sons.

Mädler, Johann H. (1873). *Geschichte der Himmelskunde von der ältesten bis aus der neueste Zeit.* Braunschweig: G. Westermann.

Mairan, Jean Jacques d'Ortous de (1744). "Sur un satellite apperçu auprès de la planète de Vénus," *Histoire de l'Académie Royale des Sciences* (for 1741) **43**, 124–128.

— (1747–48). "Von einem Trabanten den man bey der Venus beobachtet hat," *Hamburgisches Magazin*, part 4, vol. 2, 371–378.

— (1764). "Mémoire sur le satellite vu ou présumé autour de la planète de Vénus, et sur la cause de ses courtes apparations et de ses longues disparitions," *Mémoires de l'Academie Royale des Sciences* (for 1762) **64**, 161–170.

Malcolm, Noel and Jaqueline Stedall (2005). *John Pell (1611–1685) and his Correspondence with Sir Charles Cavendish: The Mental World of an Early Modern Mathematician.* Oxford: Oxford University Press.

Mandrino, Agnese, Guido Tagliaferri and Pasquale Tucci, eds. (1986). *Catalogo della Corrispondenza degli Astronomi di Brera.* Milan: Università degli Studi di Milano.

Maraldi, Giacomo (1721). "Observations sur l'etoile changeante de la Baleine," *Mémoires de l'Academie Royale des Sciences* (for 1719) **21**, 94–103.

Martin, Benjamin (1781). *The Young Gentleman and Lady's Philosophy, in a Continued Survey of the Works of Nature and Art*, vol. 2. London: W. Owen.

Martinet, Johannes F. (1779). *Katechismus der Natuur*, vol. 4. Amsterdam: Johannes Allart.

Mattison, Hiram (1849). *An Elementary Astronomy for Academies and Schools.* New York: Huntington and Savage.

Mayer, Andreas (1762). *Observationes Veneris Gryphiswaldenses quibus adiecta est M. Lamb. Henr. Röhli Reg. observ. astron. observationis suae de transitu Veneris per Solem expositio.* Greifswald: A. F. Röse.

— (1769). "Observatio ingressus Veneris in Solem 3 die Junii, 1769, habita Gryphiswaldiæ," *Philosophical Transactions* **59**, 284–285.

Meadows, A. J. (1966). "The discovery of an atmosphere on Venus," *Annals of Science* **22**, 117–127.

Meeus, Jean (1963). "En marge d'un soi-disant satellite de Vénus: La conjonction Vénus-Uranus du 4 Mars 1764," *Ciel et Terre* **79**, 36–39.

Mersenne, Marin (1945–88). *Correspondance du P. Marin Mersenne,* 17 vols. Ed. by Cornelis de Waard. Paris: Centre National de la Recherche Scientifique.

Messier, Charles (1767). "A discovery, with observations, of two new comets in the Marine Observatory at Paris," *Philosophical Transactions* (for 1766) **56**, 57–65.

— (1784). "Mémoire contenant les observations de la comète observée à Paris," *Mémoires de l'Academie Royale des Sciences* (for 1780) **82**, 515–541.

Metz, Don (2006). "William Wales and the 1769 transit of Venus: Puzzle solving and the determination of the astronomical unit," *Science and Education* (online version).

Michaud, M., ed. (1857). *Biographie universelle, ancienne et moderne,* vol. 18 Paris: C. Desplaces.

Mikkola, Seppo et al. (2004). "Asteroid 2002 VE68, a quasi-satellite of Venus," *Monthly Notices of the Royal Astronomical Society* **351**, L63–L65.

Moesgaard, Kristian P. (1972). "Horrebow, Christian," pp. 512–513 in *Dictionary of Scientific Biography,* vol. 6. New York: Charles Scribner's Sons.

Montucla, Jean-Etienne (1802). *Histoires des mathématiques,* vol. 4. Paris: Henri Agasse.

Moore, Patrick (1956). *The Planet Venus.* London: Faber and Faber.

— (1960). "A defence of Schröter," *Journal of the British Astronomical Association* **70**, 363–368.

— (1983). *The Guinness Book of Astronomy. Facts & Feats.* Enfield: Guinness Superlatives.

— (1985). "The mapping of Venus," *Journal of the British Astronomical Association* **95**, 50–61.

Newcomb, Simon (1883). "On Hell's alleged falsification of his observations of the transit of Venus in 1769," *Monthly Notices of the Royal Astronomical Society* **43**, 371–381.

Newton, Isaac (1999). *The Principia. Mathematical Principles of Natural Philosophy.* Ed. and trans. by I. Bernard Cohen and Anne Whitman. Berkeley: University of California Press.

Nicolson, Marjorie H. (1948). *Voyages to the Moon.* New York: Macmillan.

Nielsen, Axel V. (1957a). "Christian Horrebows observationer af Venuspassagen i 1761," *Nordisk Astronomisk Tidsskrift*, 47–50.

— (1957b). "Pater Hell og Venuspassagen 1769," *Nordisk Astronomisk Tidsskrift*, 77–97.

Nieto, Michael M. (1972). *The Titius–Bode Law of Planetary Distances.* Oxford: Pergamon Press.

Nisbet, H. B. (1970). *Herder and the Philosophy and History of Science.* Cambridge: Modern Humanities Research Association.

Nordenmark, N. V. E. (1939). *Per Wilhelm Wargentin.* Uppsala: Almqvist & Wiksell.

Nye, Mary Jo (1980). "N-rays: An episode in the history and psychology of science," *Historical Studies in the Physical Sciences* **11**, 125–156.

Olbers, Heinrich W. (1804). "Ueber den ersten Kometen von 1780," *Astronomisches Jahrbuch oder Ephemeriden* (for 1801), 172–181.

Orchiston, Wayne and Derek Howse (1998). "From transit of Venus to teaching navigation: The work of William Wales," *Astronomy & Geophysics* **39**, no. 6, 21–24.

Palti, Elias (1999). "The 'metaphor of life': Herder's philosophy of history and uneven developments in late eighteenth-century natural sciences," *History and Theory* **38**, 322–347.

Pappas, John (1996). "R. J. Boscovich et l'Académie des Sciences de Paris," *Revue d'Histoire des Sciences* **49**, 401–414.

Paulian, Aimé-Henri, ed. (1781). *Dictionnaire de physique, dédié au roi.* Paris: Gaude, Pere, Fils & Compagnie.

Pedersen, Olaf (1992). *Lovers of Learning: A History of the Royal Danish Academy of Sciences and Letters 1742–1992.* Copenhagen: Munksgaard.

Perkins, Jean A. (1965). "Voltaire and the natural sciences," *Studies on Voltaire and the Eighteenth Century* **37**, 61–76.

Petit, Fréderic (1846). "Sur le bolide du 21 mars 1846, et sur les conséquences qui sembleraient devoir résulter de son apparation," *Comptes Rendus des Séances de l'Académie des Sciences* **23**, 704–709.

— (1847). "Sur le bolide du 23 juillet 1846," *Comptes Rendus des Séances de l'Académie des Sciences* **25**, 259–262.

Pickering, Edward C. and William H. Pickering (1890). "Photographic search for a lunar satellite," *Annals of Harvard College Observatory* **18**, 77–83.

Pickering, William H. (1923). "A meteoric satellite," *Popular Astronomy* **31**, 82–85.

Pingré, Alexandre-Guy (1763). "Observation of the transit of Venus over the Sun, June 6, 1761," *Philosophical Transactions* (for 1761) **52**, 371–377.

Plassmann, Joseph (1898). *Himmelskunde.* Freiburg: Herder.

Poggendorff, J. C. (1863). *Biographisch-literarisches Handwörterbuch zur Geschichte der exacten Wissenschaften.* Leipzig: J. A. Barth.

Popper, Karl R. (1972). *Conjectures and Refutations: The Growth of Scientific Knowledge.* London: Routledge and Kegan Paul.

Priestley, Joseph (1772). *The History and Present State of Discoveries Relating to Vision, Light, and Colours.* London: J. Johnson.

Prince, C. Leeson (1882). "Observation of the transit of Venus, December 6, made at Crowborough, Sussex," *Monthly Notices of the Royal Astronomical Society* **43**, 64–65.

— (1883). *Observations upon the late Great Comet and Transit of Venus, made at Crowborough, Sussex, in the Year 1882.* Lewes: H. W. Wolff.

Proctor, Richard A. (1890). *Other Worlds than Ours.* New York: Appleton.

Ranyard, Arthur C. (1889). "Richard Anthony Proctor," *Monthly Notices of the Royal Astronomical Society* **49**, 164–168.

Riccioli, Giambattista (1651). *Almagestum novum astronomiam veterem novamqve complectens.* Bologna: Haeredis Victorii Benatii.

Roe, Shirley A. (1985). "Voltaire versus Needham: Atheism, materialism, and the generation of life," *Journal of History of Ideas* **46**, 65–87.

Roedkiær, Peder (1765). "Beretning om én den 3 og 4 Martii 1764 giort iagttagelse angaaende Veneris drabant," *Skrifter, Kgl. danske Videnskabernes Selskab,* Part 9 (for 1761–64), 394–395.

Röhl, Lampert H. (1768). *Merkwürdigkeiten von den Durchgängen der Venus durch die Sonne.* Greifswald: A. F.

Rosen, Edward, ed. (1959). *Three Copernican Treatises.* New York: Dover Publications.

Roseveare, N. T. (1982). *Mercury's Perihelion. From Le Verrier to Einstein.* Oxford: Clarendon Press.

Russell, Henry N., Raymond S. Dugan and John Q. Stewart (1926). *Astronomy.* Boston: Ginn and Company.

Sarton, George (1936). "Montucla (1725–1799): His life and works," *Osiris* **1**, 519–567.

— (1944). "Vindication of Father Hell," *Isis* **35**, 97–105.

Schaefer, Bradley E. (2001). "The transit of Venus and the notorious black drop effect," *Journal for the History of Astronomy* **32**, 325–336.

Schaffer, Simon (1981). "Uranus and the establishment of Herschel's astronomy," *Journal for the History of Astronomy* **12**, 11–26.

Schjellerup, Hans C. F. C. (1882). "On some hitherto unknown observations of a supposed satellite of Venus," *Copernicus: An International Journal of Astronomy* **2**, 164–168.

Schlyter, Paul (2003). "Hypothetical worlds – Earth's second moon," *Arcturus* **1**, 4–6.

Schorr, F. (1873). *Der Vorübergang der Venus von der Sonnenscheibe am 9. December 1874 und die Bestimmung der Entfernung der Sonne.* Braunschweig: Vieweg und Sohn.

— (1875). *Der Venusmond und die Untersuchungen über die früheren Beobachtungen dieses Mondes.* Braunschweig: Vieweg und Sohn.

Schröter, Johann H. (1792). "Observations on the atmospheres of Venus and the Moon, their respective densities, perpendicular heights, and the twilight occasioned by them," *Philosophical Transactions* **82**, 309–361.

— (1796). *Aphroditographische Fragmente, zur genauern Kenntnis des Planeten Venus.* Helmstedt: C. G. Fleckeisen.

— (1798). "Observations on the satellites of the Georgium planet, and the two innermost satellites of Saturn," *Philosophical Magazine* **2**, 282–290.

— (1995). *Beobachtungen über die sehr beträchtlichen Gebirge und Rotation der Venus.* Ed. by Dieter Gerdes. Lilienthal: Verlag M. Simmering.

See, Thomas J. J. (1909). "Dynamical theory of the capture of satellites and of the division of nebulae under the secular action of a resisting medium," *Astronomische Nachrichten* **181**, columns 333–350.

Senebier, Jean (1802). *Essai sur l'art d'observer et de faire des expériences.* Geneva: J. J. Paschoud.

Shea, William R. (1970). "Galileo, Scheiner, and the interpretation of Sunspots," *Isis* **61**, 498–519.

Sheehan, William (1988). *Planets & Perceptions. Telescopic Views and Interpretations, 1609–1909.* Tuczon: University of Arizona Press.

— (2002). "The tragic case of T. J. J. See," *Mercury* **31**, no. 6, 34–39.

Sheehan, William and Thomas Dobbins (2003). "The spokes of Venus: An illusion explained," *Journal for the History of Astronomy* **34**, 53–63.

Sheehan, William and John E. Westfall (2004). *The Transits of Venus.* Amherst: Prometheus Books.

Shklovskii, Iosef and Carl Sagan (1966). *Intelligent Life in the Universe.* New York: Dell.

Short, James (1744). "An observation on the planet Venus (with regard to her having a satellite)," *Philosophical Transactions* **41** (for 1739–41), no. 459, 646–647.

— (1762). "An account of the transit of Venus over the Sun, on Saturday morning, 6th June 1761," *Philosophical Transactions* (for 1761) **52**, 178–182.

Smith, Robert W. and Richard Baum (1984). "William Lassell and the ring of Neptune: A case study in instrumental failure," *Journal for the History of Astronomy* **15**, 1–17.

Smye-Rumsby, Greg (2004). "London's lost Leviathan," *Astronomy Now* **18**, no. 10, 33–36.

Smyth, William H. (1844). *A Cycle of Celestial Objects*, 2 vols. London: John W. Parker.

Spencer, Herbert (1858). "Recent astronomy, and the nebular hypothesis," *Eclectic Magazine of Foreign Literature, Science, and Art*, September–December, 73–87.

— (1891). *Essays: Scientific, Political, and Speculative*, vol. 1. London: Williams and Norgate.

Steel, Duncan (1982). "The Craig telescope of 1852," *Sky and Telescope* **64**, no. 7, 12–13.

Stroobant, Paul (1887a). "Études sur le satellite énigmatique de Vénus," *Mémoires Couronnés et Mémoires des Savants Étrangers Publiés par l'Académie Royale des Sciences, des Lettres et des Beaux-Arts de Belgique*, vol. 49, no. 5 (48 pp.).

— (1887b). "Le satellite énigmatique de Vénus," *L'Astronomie* **6**, 452–457.

— (1887c). "Le satellite énigmatique de Vénus," *La Nature* **15**, 307–308.

— (1888). "Étude sur le satellite énigmatique de Vénus," *Astronomische Nachrichten* **118**, no. 2809, columns 5–10.

Swift, Jonathan (1994). *Gulliver's Travels.* London: Penguin Books.

Tacquet, Andreas (1669). *Opera mathematica.* Antwerp: Jacobus Meur.

Taton, René (1984). "D'Alembert, Euler et l'Académie de Berlin," *Dix-Huitième Siècle* **16**, 55–69.

— (2000). *Etudes d'histoire des sciences.* Ed. by Danielle Fauque, Myriana Ilic and Robert Halleux. Turnhout: Brepols

Terrall, Mary (1990). "The culture of science in Frederick the Great's Berlin," *History of Science* **28**, 333–364.

Thirion, J. (1885). "Le satellite de Vénus," *Revue des Questions Scientifique* **17**, 44–62.

Thornton, Robert J. (1803). *A New Illustration of the Sexual System of Linnæus.* London: T. Bensley.

— (1804). "Letter to Mr. Arthur Aikin from Mr. Thornton," *Philosophical Magazine* **18**, 326–332.

Torricelli, Evangelista (1919). *Opere di Evangelista Torricelli*, vol. 3. Ed. by Gino Loria and Giuseppe Vassura. Faenza: G. Montanari.

Turner, Gerard L'E. (1969). "James Short, F.R.S., and his contributions to the construction of reflecting telescopes," *Notes and Records of the Royal Society of London* **24**, 91–108.

Tuttle, Hudson and Emmet Denmore (2003). *Arcana of Nature.* New York: Kessinger.

Tuxen, Johan C. (1861). *Sol- eller Planetsystemet. Verdenslegemernes Bevægelser, deres Stilling i Universet og Physiske Forhold.* Copenhagen: Philipsens Forlag.

Van Flandern, Thomas C. and Robert S. Harrington (1976). "A dynamical investigation of the conjecture that Mercury is an escaped satellite of Venus," *Icarus* **28**, 435–440.

Van der Wall, Ernestine G. E. (2004). "Newtonianism and religion in the Netherlands," *Studies in History and Philosophy of Science* **35**, 493–514.

Van Helden, Albert (1976a). "The 'astronomical telescope' 1611–1650," *Annali dell'Istituto e Museo di Storia della Scienza* **1**, 13–36.

— (1976b). "The importance of the transit of Mercury of 1632," *Journal for the History of Astronomy* **7**, 1–10.

— (1994). "Telescopes and authority from Galileo to Cassini," *Osiris* **9**, 7–29.

Verhas, Pierre (2002). *Jean-Charles Houzeau et son temps.* Brussels: Arthur Mergelynck Foundation.

Vogel, H. C., ed. (1905), *Newcomb-Engelmanns Populäre Astronomie*, 3rd edn. Leipzig: Verlag W. Engelmann.

Voiron, M. (1810). *Histoire de l'astronomie depuis 1781 jusqu'a 1811.* Paris: V. Courcier.

Voltaire (2000). *Micromégas.* Ed. and annotated by Jean Goulemot. Paris: Livre de Poche.

Wales, William and Joseph Dymond (1770). "Astronomical observations made by order of the Royal Society, at Prince of Wale's Fort, on the north-west coast of Hudson's Bay," *Philosophical Transactions of the Royal Society* (for 1769) **59**, 467–488.

Wallis, Ruth (1982). "John Bevis, M.D., F.R.S. (1695–1771): Astronomer loyal," *Notes and Records of the Royal Society of London* **36**, 211–225.

Waltemath, Georg (1898). *Aufruf an die Herren Astronomen und Freunde der Astronomie: Ein zweiter Mond der Erde.* Hamburg.

Ward, William R. and Mark J. Reid (1973). "Solar tidal friction and satellite loss," *Monthly Notices of the Royal Astronomical Society* **164**, 21–32.

Wargentin, Per W. (1780). "De simulato satellite Veneris monitum," *Nova Acta Regiæ Societatis Scientiarum Upsaliensis* **3**, 224.

Warner, Brian (1977). "W. E. Wilson and the Daramona Observatory," *Sky and Telescope* **53**, 108–110.

Watts, George B. (1958). "The Encyclopédie méthodique," *PMLA* (Proceedings of the Modern Language Association) **73**, 348–366.

Webb, Thomas W. (1859). *Celestial Objects for Common Telescopes.* London: Longman, Green, Longman, and Roberts.

— (1868). "Supposed satellite of Venus," *Astronomical Register* **6**, 224.

— (1876). "The satellite of Venus," *Nature* **14**, 193–195.

Whewell, William (1856). *Astronomy and General Physics Considered with Reference to Natural Theology.* New York: Harper & Brothers.

Willach, Rolf (2001). "The development of telescope optics in the middle of the seventeenth century," *Annals of Science* **58**, 381–398.

Wilson, H. C. (1887). "The supposed satellite of Venus," *The Sidereal Messenger* **6**, 357–359.

Winkler, Mary G. and Albert van Helden (1992). "Representing the heavens: Galileo and visual astronomy," *Isis* **83**, 195–217.

— (1993). "Johannes Hevelius and the visual language of astronomy," pp. 97–116 in J. V. Field and Frank A. J. L. James, eds., *Renaissance and Revolution: Humanists, Scholars, Craftmen and Natural Philosophers in Early Modern Europe.* Cambridge: Cambridge University Press.

Winthrop, John (1764). "Observations of the transit of Venus, June 6, 1761," *Philosophical Transactions* (for 1764) **54**, 279–283.

Woeckel, Johann S. L. (1846). *Die Sonne und ihre Flecken.* Nuremberg: F. Campe.

Wolf, Charles J. E. (1902). *Histoire de l'Observatoire de Paris de sa fondation a 1793.* Paris: Gauthier-Villars.

Wolf, J. Rudolf (1857). "Mittheilungen über die Sonnenflecken," *Vierteljahrschrift der naturforschenden Gesellschaft in Zürich* **2**, 109–132, 272–299, 349–395.

— (1877). *Geschichte der Astronomie.* Munich: R. Oldenbourg.

— (1891). *Handbuch der Astronomie. Ihrer Geschichte und Litteratur*, 2 vols. Zürich: F. Schultess.

Wolff, Christian (1752). *Vernünfftige Gedancken von den Absichten der natürlichen Dinge.* Halle: Rengerischen Buchhandlung.

Woolf, Harry (1959). *The Transits of Venus. A Study of Eighteenth-Century Science.* Princeton: Princeton University Press.

Wurm, Johann F. (1802). "Versuch einer genaueren Bestimmung der Massen der Planeten," (Zach's) *Monatliche Correspondenz zur Beförderung der Erd- und Himmelkunde* **5**, 546–570.

Yeo, Richard (1986). "The principle of plenitude and natural theology in nineteenth-century Britain," *British Journal for History of Science* **19**, 263–282.

Young, Charles A. (1886). "The year's progress in astronomy," *Transactions of the New York Academy of Sciences* **5**, 234–263.

— (1893). *A Text-Book of General Astronomy.* Boston: Ginn and Company.

Zahn, Johann (1702). *Oculus artificialis teledioptricus sive telescopium.* Nuremberg: Johannis C. Lochneri.

Zenger, Carl V. (1877). "Absorption of the light of Venus by dark violet glass plates," *Monthly Notices of the Royal Astronomical Society* **37**, 460–462.

Index

Science Networks – Historical Studies (SNHS)

Edited by
Eberhard Knobloch, Technische Universität Berlin, Germany
Helge Kragh, University of Aarhus, Denmark
Erhard Scholz, Bergische Universität Wuppertal, Germany

In cooperation with an international editorial board

BIRKHÄUSER

The publications in this series are limited to the fields of mathematics, physics, astronomy, and their applications. The publication language is preferentially English. The series is primarily designed to publish monographs. Annotated sources and exceptional biographies might be accepted in rare cases. The series is aimed primarily at historians of science and libraries; it should also appeal to interested specialists, students, and diploma and doctoral candidates. In cooperation with their international editorial board, the editors hope to place a unique publication at the disposal of science historians throughout the world.

SNHS 37: Kragh, H.
The Moon that Wasn't. The Saga of Venus' Spurious Satellite (2008)
ISBN 978-3-7643-8908-6

SNHS 36: Kvasz, L.
Patterns of Change. Linguistic Innovations in the Development of Classical Mathematics (2008)
ISBN 978-3-7643-8839-3

This book offers a reconstruction of linguistic innovations in the history of mathematics; innovations which changed the ways in which mathematics was done, understood and philosophically interpreted. It argues that there are at least three ways in which the language of mathematics has been changed throughout its history, thus determining the lines of development that mathematics has followed.
One of these patterns of change, called a re-coding, generates two developmental lines. The first of them connecting arithmetic, algebra, differential and integral calculus and predicate calculus led to a gradual increase of the power of our calculating tools, turning difficult problems of the past into easy exercises. The second developmental line connecting synthetic geometry, analytic geometry, fractal geometry, and set theory led to a sophistication of the ways we construct geometrical objects, altering our perception of form and increasing our sensitivity to complex visual patterns.
Another important pattern of change, called relativization, is illustrated by the development of synthetic geometry, connecting Euclid's geometry, projective geometry, non-Euclidean geometry, and Klein's Erlanger Programm up to Hilbert's Grundlagen der Geometrie. In this development the notions of space and geometric object underwent deep and radical changes culminating in the liberation of objects from the supremacy of space and so bringing to existence geometric objects which space would never tolerate.

SNHS 35: Caramalho Domingues, J.
Lacroix and the Calculus (2008)
ISBN 978-3-7643-8637-5

Silvestre François Lacroix (Paris, 1765 – ibid., 1843) was a most influential mathematical book author. His most famous work is the three-volume *Traité du calcul différentiel et du calcul intégral* — an encyclopedic appraisal of 18th-century calculus which remained the standard reference on the subject through much of the 19th century, in spite of Cauchy's reform of the subject in the 1820's.
Lacroix and the Calculus is the first major study of Lacroix's large *Traité*. It uses the unique and massive bibliography given by Lacroix to explore late 18th-century calculus, and the way it is reflected in Lacroix's account. Several particular aspects are addressed in detail, including: the foundations of differential calculus, analytic and differential geometry, conceptions of the integral, and types of solutions of differential equations (singular/complete/general integrals, geometrical interpretations, and generality of arbitrary functions).

SNHS 34: Høyrup, J.
Jacopo da Firenze's *Tractatus Algorismi* and Early Italian Abbacus Culture (2007)
ISBN 978-3-7643-8390-9

Science Networks – Historical Studies (SNHS)

Edited by
Eberhard Knobloch, Technische Universität Berlin, Germany
Helge Kragh, University of Aarhus, Denmark
Erhard Scholz, Bergische Universität Wuppertal, Germany
In cooperation with an international editorial board

BIRKHÄUSER

SNHS 33: De Risi, V.
Geometry and Monadology. Leibniz's *Analysis Situs* and Philosophy of Space (2007)
ISBN 978-3-7643-7985-8

SNHS 32: Krömer, R.
Tool and Object. A History and Philosophy of Category Theory (2007)
ISBN 978-3-7643-7523-2

SNHS 31: Keller, A.
Expounding the Mathematical Seed. Vol. 2: The Supplements (2006). ISBN 978-3-7643-7292-7

SNHS 30: Keller, A.
Expounding the Mathematical Seed. Vol. 1: The Translation (2006). ISBN 978-3-7643-7291-0

SNHS 30/31 Set: ISBN 978-3-7643-7299-6

SNHS 29: Guerraggio, A. / Nastasi, P.
Italian Mathematics Between the Two World Wars (2005). ISBN 3-7643-6555-2

SNHS 28: Hesseling, D.
Gnomes in the Fog. The Reception of Brouwer's Intuitionism in the 1920s (2003)
ISBN 978-3-7643-6536-3

SNHS 27: Dauben, J.W. / Scriba, C.J.
Writing the History of Mathematics – Its Historical Development (2002)
ISBN 978-3-7643-6166-2 (Hardcover)
ISBN 978-3-7643-6167-9 (Softcover)

SNHS 26: Israel, G. / Millán Gasca, A.
The Biology of Numbers. The Correspondence of Vito Volterra on Mathematical Biology (2002)
ISBN 978-3-7643-6514-1

SNHS 25: Siegmund-Schultze, R.
Rockefeller and the Internationalization of Mathematics Between the Two World Wars. (2001). ISBN 978-3-7643-6468-7

SNHS 24: Jensen, C.
Controversy and Consensus: Nuclear Beta Decay 1911–1934. (2000)
ISBN 978-3-7643-5313-1

SNHS 23: Ferreirós, J.
Labyrinth of Thought. A History of Set Theory and its Role in Modern Mathematics (2001)
ISBN 978-3-7643-5749-8

SNHS 22: Marage, P. / Wallenborn, G. (eds.)
The Solvay Councils and the Birth of Modern Physics (1999). ISBN 978-3-7643-5705-4

SNHS 21: Sakarovitch, J.
Épures d'architecture. De la coupe des pierres à la géométrie descriptive XVI–XIX siècles (1998)
ISBN 978-3-7643-5701-6

SNHS 20: Grattan-Guinness, I. / Bornet, G.
George Boole – Selected Manuscripts on Logic and its Philosophy (1997)
ISBN 978-3-7643-5456-5

SNHS 19: Ullmann, D.
Chladni und die Entwicklung der Akustik 1750–1860 (1996). ISBN 978-3-7643-5398-8

SNHS 18: Hentschel, K. (ed.)
Physics and National Socialism. An Anthology of Primary Sources (1996)
ISBN 978-3-7643-5312-4

SNHS 17: Corry, L.
Modern Algebra and the Rise of Mathematical Structures (1996). ISBN 978-3-7643-5311-7

SNHS 16: Yavetz, I.
From Obscurity to Enigma. The Work of Oliver Heaviside, 1872–1889 (1995)
ISBN 978-3-7643-5180-9

SNHS 15: Sasaki, Ch. / Sugiura, M. / Dauben, J.W., The Intersection of History and Mathematics (1994). ISBN 978-3-7643-5029-1

MIX
Papier aus verantwortungsvollen Quellen
Paper from responsible sources
FSC® C105338

If you have any concerns about our products,
you can contact us on
ProductSafety@springernature.com

In case Publisher is established outside the EU,
the EU authorized representative is:
Springer Nature Customer Service Center GmbH
Europaplatz 3, 69115 Heidelberg, Germany

Printed by Libri Plureos GmbH
in Hamburg, Germany